The White Paper

The White Paper

by Satoshi Nakamoto

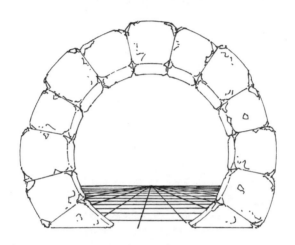

Introduction by
James Bridle

Edited by
Jaya Klara Brekke and Ben Vickers

First published by Ignota 2019
This selection © Ignota 2019

'Bitcoin: A Peer-to-Peer Electronic Cash System'
© Satoshi Nakamoto 2008
Excerpts from the Cryptography Mailing List © the individual contributors
Excerpts from Bitcointalk forum © the individual contributors
Excerpts from the Cypherpunks Mailing List (cypherpunks@toad.com)
© the individual contributors

1 3 5 7 9 10 8 6 4 2

Ignota
ignota.org

ISBN-13: 978-1-9996759-2-9

Design by Cecilia Serafini
Cover detail: The Ambassadors (1533)
by Hans Holbein the Younger
Typeset in Freight Text and Acumin Pro
Printed and bound in Great Britain by TJ International

Contents

F

Preface

by Ben Vickers

This book is a strange but important undertaking, produced with specific intent. It has been written and compiled for the sceptics, the concerned and the still curious. It seeks to do nothing more than make legible, in simple terms, a system of knowledge that, in its current form or another, has profound implications for how human beings might choose to organise their societies over the course of the next century.

In 2008, a figure known only by the alias Satoshi Nakamoto appeared on *The Cryptography Mailing List* to share a short and astonishingly concise technical paper entitled *Bitcoin: A Peer-to-Peer Electronic Cash System*. Proposing 'a new electronic cash system that's fully peer-to-peer, with no trusted third party', its publication in the wake of the global financial crisis was both prescient and profound. In a mere nine pages, Nakamoto's white paper presented a technically robust set of solutions for fundamental challenges that had, for decades, thwarted the aspirations of the Cypherpunk and crypto-anarchist movements. Their utopian vision of a decentralised, uncensorable and free internet had, for many, died prematurely in the commercialisation and subsequent crash of the dot-com era, and with the more recent enclosures of access and virtual territory orchestrated by the Big Five: Google, Amazon, Facebook, Apple and Microsoft.

The first iteration of the Bitcoin protocol was deployed into the world in January 2009. This publication marks its tenth anniversary with a moment of deep reflection and contemplation. In the decade that has passed, and given the radically alien nature of the technology, its success has been significant. Following the disappearance of Satoshi Nakamoto in or around 2011, the dominant media narratives and the plethora of articles written on the subject have tended towards two concerns: first, the identity of Satoshi Nakamoto and second, the high volatility of cryptocurrency, producing a market value that at its peak operated in excess of $200 billion. It should then be made clear from the outset that this book is not concerned with either of these questions; rather, this is a journey through the deeply complex set of challenges that have arisen out of such success, and the poignant confrontation it presents to existing systems of governance, both technical and social.

Origin myths shrouded in mystery, and seemingly limitless capital growth will inevitably garner significant attention in our current age – but stories that focus on either of these are narratives of speculation which heavily obscure, polarise and distract us from the more seismic implications for what is now in play as a result of the adoption of blockchain technologies.

Instead, this is a book about the threads of history that have woven its path, as laid out in James Bridle's elegant and historical analysis of the blockchain's place in the world, as well as the new and emergent political spectrum that has been catalysed as a result of embedding increasingly advanced technologies such as blockchain into our everyday infrastructures. The complexities and choices made by those engineering such a future are discussed in Jaya Klara Brekke's exhaustive and complete excavation of the socio-technical doctrine that constitutes the genesis of Satoshi Nakamoto's creation and all that has followed. This deep analysis of blockchain's inception in the white paper helps to trace the lines, antagonisms and connections between the current architecture for the systems of power that are seeking to maintain centralised control, and those competing ascendent forces manifest through Cypherpunk-borne technologies and ideologies that seek to disburse and decentralise such power as we know it. Brekke walks a sensitive and considerate path that favours neither, but which instead lays the ground for careful thinking between the two. With some of the hindsight wrought by a decade of development, it has become self-evident that these shifts cannot, by default, enable either better or worse conditions for any single individual, group or body politic. This book, which is both guide and commentary, allows us to become witness and participant to the redrawing of the maps of political possibility – and of an emergent political reality that continues to unfold in real time. In doing so, it presents a set of options and long-term trajectories that can only be steered by those who choose to participate.

And for those architects and engineers who have boldly set out to construct this new world, we urge you to pay close attention to the nuance with which the basic assumptions of Satoshi's work have been extrapolated and unpicked in this book. It is a necessary and generous confrontation with the reality of the world you are constructing.

. . .

Lastly, and to make sense of the context in which this publication came to be the second book published by Ignota Books: Arthur C. Clarke once remarked, 'Any sufficiently advanced technology is indistinguishable from magic.' It should be clear to those paying close attention that the magical texts of now, which act to radically transform the world, will rarely reveal themselves as such within the era of their making. But give consideration to the situation; it's not since the early formation of the Christian Church that a text or doctrine handed down without claim to authority has produced such extreme sectarian conflict and zealotry as Satoshi Nakamoto's white paper.

Introduction

by James Bridle

It's difficult to know when humans first started securing or 'encrypting' messages to hide them from unwanted readers; the practice must, by human nature, be almost as old as written language, although examples are sparse. We know, for example, that Julius Caesar used a simple form of letter substitution to communicate with his generals, shifting each character three steps down the alphabet in order to scramble it. The ancient Greeks, particularly the military-minded Spartans, used a device called a *scytale*, which allowed a hidden text to be read by wrapping a strip of parchment around a cylinder of a particular size so that the letters lined up in a particular order.

Tales of the Greco-Persian Wars are full of secret messages, not least the story of Histiaeus, a commander who, according to Herodotus, shaved the head of his favourite slave and had it tattooed with a message urging revolution in the city of Miletus. When the slave's hair grew back he was dispatched to the city, with the instructions that the recipient should shave him once again and read the message there revealed. Such extreme measures were taken due to the fear of government surveillance, a justification often cited today. The Persian king controlled the roadways, and had the power to examine any message – and messenger – that travelled on them. From the very beginning, cryptography has been both a military technology and a tool for undermining existing powers.

Cryptography's value as a military tool is double-edged, of course. Like other weapons, its effectiveness depends on the ability of one side to outgun the other. For a long time, this balance mostly held, with efforts by one side to crack the secrets of the other forming long-running and fascinating backstories to many conventional conflicts. It was an act of decryption that brought the United States into the First World War, when British intelligence services decoded the infamous Zimmermann Telegram proposing an alliance between Germany and Mexico. In the closing months of the war, the cracking of Germany's ADFGVX cipher by French cryptanalysts enabled the Allies to stave off a final German offensive on Paris.

Cryptography was first mass-manufactured in the Second World War, in the form of the Third Reich's Enigma machines, and then digitised in the form of the Colossus, the world's first programmable electronic computer, developed to break the German military's Lorenz cipher. The wild invention and ultimate success of the Bletchley code-breakers over their Nazi adversaries can be read as the first of many instances of the digital overcoming the physical; the Lorenz SZ42 was a massive, complex machine of rotating cogs and wheels which defied codebreakers for years. By the end of the war, it was completely readable by an electronic machine. The secrecy around the Colossus itself meant that its existence had little influence on future computer design, but it marks the point at which cryptography changed radically in nature – because what is digital is ultimately distributable, although it would take the growth of the internet in the 1990s for this to become widely understood.

In 1991, a computer security researcher called Phil Zimmermann created a programme called Pretty Good Privacy (PGP), which enabled users of home computers to strongly encrypt email messages using a combination of numerous well-known algorithms. What turned PGP from another home made software product into one of the most contentious artefacts of the decade wasn't how it was made, but how it was distributed. Since the Second World War, nations had been forced to legally define cryptography as a weapon; like any other munition, cryptography was subject to something called the Arms Export Control Act. At the time of PGP's release, any cryptosystem which used keys – the strings of randomly generated numbers which secured hidden messages – longer than 40 bits required a licence for export. PGP used keys which were 128 bits long and almost impossible to crack at the

time, and this made it precisely the kind of technology that US authorities wanted to prevent falling into foreign hands. Zimmermann never
intended to export PGP, but, fearing that it would be banned outright,
he started distributing it to friends, saying, 'I wanted to strengthen
democracy, to ensure that Americans could continue to protect their
privacy.'[1] Shortly after that, PGP found its way onto the internet and
then abroad. In 1993, the US government started a formal investigation
into Zimmermann – for exporting munitions without a license.

As knowledge of the case spread, it became a flashpoint for early
digital activists who insisted on the rights of everyone to protect their
own secrets and their own private lives. The freedoms and dangers of
code became the subject of earnest debate, and in another foreshad-
owing of future digital style, of hacks, pranks and stunts. Zimmermann
had the software's source code printed as a hardback book, allowing
anyone to purchase a copy and type up the software themselves. As
he was fond of pointing out, export of products commonly considered
munitions – bombs, guns and planes – could be restricted, but books
were protected by the First Amendment. Variants on the RSA algorithm
– the 128-bit process at the heart of PGP – were printed on T-shirts
bearing the message 'This shirt is classified as a munition.' Some went
further, having lines of code tattooed onto their arms and chests.

The Crypto Wars, as they became known, galvanised a community
around the notion of personal – rather than national – security, which
tied into the utopian imagination of a new, more free, more equal and
more just society developing in cyberspace. Another development
that prompted widespread public disquiet was the US government's
proposal for a chipset for mobile phones. The Clipper chip was designed
by the NSA to provide encryption for users while allowing law enforcement to eavesdrop on communications – a situation that was ripe for
abuse, either by corrupt officials or by skilled hackers. The idea that a

1. 'Lost in Kafka Territory', *US News*, March 1995, https://web.archive.org/web/20130616165334/
    http://www.usnews.com/usnews/news/articles/950403/archive_010975.htm

government would deliberately weaken the protections available to its citizens made for an even more powerful and accessible argument for the individualists than the attack on PGP. By the late nineties, Clipper was dead – and so was the case against Zimmermann. The hackers and privacy activists declared victory in the Crypto Wars. Yet what's often regarded as a victory for everyone against government overreach can also be read as a moment of terrifying breach: when the state's most powerful weapons escaped government control and fell into the hands of anyone who wanted to use them.

Today, thanks to the rise in digital communications, cryptography is everywhere, not least in banking systems, protecting the billions of electronic transactions that flow around the planet every day. Even more than in the nineties, the idea that anyone would deliberately make it easier for someone to steal money seems like an attack on the basic functions of society, and so it should come as no surprise that it's a technology best known for – but by no means limited to – the distribution of currency that should be the focus of a new outbreak of the Crypto Wars, as well as the full flood of individualist, utopian thinking that accompanied the first round. There's something about money that focuses the mind.

When Marco Polo first encountered paper money on his travels to China in the thirteenth century, he was astounded. In his *Book of the Marvels of the World*, he spends a great length of time explaining, and wondering at, the monetary system established by the Great Khan. Until recently, and as was still the case in Europe, the Chinese had used a range of value-bearing commodities to settle commerce and taxation: copper ingots, iron bars, gold coins, pearls, salt and the like. In 1260, Kublai Khan decreed that instead, his subjects would use apparently valueless paper, printed and certified by a central mint, and, writes Polo, 'the way it is wrought is such that you might say he has the secret of alchemy in perfection, and you would be right.' Through a carefully choreographed process of manufacture, design and official imprimatur, 'all these pieces of paper are issued with as much solemnity and authority as if they were of pure gold or silver! The process was alchemical in the truest sense, as it did not merely transform material, but also elevated the Khan himself to even more unassailable heights of power: the only arbiter of finance. Those who refused to accept the new currency were punished with death, and all trade flowed through the state's coffers. Like the Persian king before him, the Khan had realised that controlling traffic – in commerce and in information – was the way to situate oneself at the true heart of power.

The processing and accounting of money – fiat money, created by decree rather than having inherent value – is essentially the manipulation of symbols, and the gradual but ever-accelerating authority of capitalism, money's belief system, tracks the development of symbol-manipulating technologies, from the double-entry bookkeeping of the European Renaissance to the development of databases and planet-spanning electronic networks; from physical technologies to virtual ones.

Money also involves the magical transformation of symbols into value. It requires belief to operate. Around such belief systems other beliefs tend to gather, and the industrial quantities of belief required to breathe life into new systems of value tends to gives succour to any number of outlandish ideas, whether these be the divine right of kings, the supremacy of the nation state or the inviolable will of technology itself.

Money, then, is a belief system backed by state infrastructure which, for a long time, assured centralised power. But as computational technologies, long the sole province of the state, became less about asserting government power than asserting individual freedom – in other words, as the weapons forged in the crucible of the Second World War became increasingly available to the common citizen – it became clear to the veterans of the Crypto Wars how they might make other adjustments to ancient power dynamics.

The idea for digital money and virtual currencies had been floating around for some time before the Crypto Wars. Money has been tending towards the virtual for some time, from the first ATMs and cards in the sixties, to the spread of digital networks and connections between retailers and banks in the eighties and nineties. For anyone with a little technological foresight, it was easy to see which direction we were heading in. For those concerned with privacy and individual sovereignty, it was a worrying development. Digital money, significantly, has none of the advantages of cash; it can't be stored and exchanged outside of the system of banks and third parties, such as credit card companies, which can regulate and impede its flow. To a cryptographer, or anyone who has imbibed cryptography's lessons on the potential to separate oneself from overbearing powers, this arrangement looks like a kind of enslavement. So what would digital cash actually look like?

The first quality of digital cash is that it needs to be private, in the sense that no one other than the spender and receiver should be party to the transaction: no bank or security agency should know who is spending the money, who is receiving it, what it is for or at what time and place the exchange is taking place. Because no physical assets, such as notes or coins, are being exchanged, it should also be secure. The receiver should be able to verify they were paid and the spender that they had paid – a two-way receipt for the transaction. In this way, digital cash would have all the privacy of physical cash, with the added benefit of the participants being able to prove that a transaction had actually taken place.

One of the earliest proponents of digital cash was an American computer scientist called David Chaum. Chaum believed that both the privacy and the security problems of digital currencies could be solved by using cryptography: encoding messages between the two parties, the sender and the receiver, in such a way that nobody else can read them. Chaum's solution to this problem involved both parties digitally signing the transaction with a private key, akin to an unforgeable and unguessable digital signature. In this way, both parties validate the transaction. In addition, they communicate through encrypted channels, so that nobody else can see that the transaction has taken place.[2]

Chaum's system worked, and was implemented by a number of companies and even one bank, but it never took off. Chaum's own company, DigiCash, went bankrupt in 1998 and there was little incentive to compete against the growing power of credit card companies. Chaum felt that people didn't understand what they were losing as digital networks and the money that flowed across them became more centralised: 'As the web grew, the average level of sophistication of users dropped. It was hard to explain the importance of privacy to them', he said in 1999.[3]

Yet some people, including those radicalised by the Crypto Wars of the early nineties, did understand the value of privacy. A group which came to be known as the Cypherpunks gathered first in San Francisco, and then online, with the intent of picking up from Chaum's work the tools that could be used to disempower governments. From the very beginning, Chaum's ideas about privacy and security had been tied to ideas about society and the way it was being changed by digitisation. 'Computerisation is robbing individuals of the ability to monitor and control the ways information about them is used', he wrote in 1985, foreseeing a Big Brother-like 'dossier society' where everything was known about individuals but individuals knew little about the information held over them.[4] Yet Chaum was forced to partner with existing institutions to get DigiCash off the ground – and this was very far from the Cypherpunk dream.

Eric Hughes, a Berkeley mathematician and one of the original Cypherpunks group, published *A Cypherpunk's Manifesto* in 1993, arguing that privacy was a requirement for an open society, and privacy on electronic networks could only be achieved through the

2. These two processes are called 'blind signatures' and 'mix networks' and were published in the papers 'Blind signatures for untraceable payments' (David Chaum, 1982) and 'Untraceable electronic mail, return addresses, and digital pseudonyms' (David Chaum, 1981).
3. 'Requiem for a Bright Idea', *Forbes Magazine*, 1 November 1999
4. 'Security without Identification: Transaction Systems to Make Big Brother Obsolete', David Chaum, 1985

use of cryptography.[5] Tim May, another member of the group and a former chief scientist at Intel, went further in the *The Crypto Anarchist Manifesto*: 'The State will of course try to slow or halt the spread of this technology, citing national security concerns, use of the techno-logy by drug dealers and tax evaders and fears of societal disintegra-tion. Many of these concerns will be valid; crypto anarchy will allow national secrets to be traded freely and will allow illicit and stolen materials to be traded. An anonymous computerised market will even make possible abhorrent markets for assassinations and extortion. Various criminal and foreign elements will be active users of CryptoNet. But this will not halt the spread of crypto anarchy. Just as the techno-logy of printing altered and reduced the power of medieval guilds and the social power structure, so too will cryptologic methods fundamen-tally alter the nature of corporations and of government interference in economic transactions.'[6]

Throughout the nineties and into the noughties, the Cypherpunks elaborated on the principles that would bring their utopia of encryp-tion into being, as well as the technical innovations required to make digital currency possible. One of the biggest hurdles to doing so was the double-spending problem. Physical cash can only be spent once; when a banknote is handed over to a merchant, the buyer can't at the same time use the same note at another shop round the corner. Virtual currencies face the problem that while encryption can guarantee that this specific piece of data is a form of money belonging to this specific person, it can't say whether that data has been copied and is also in circulation elsewhere. In other words, it can't say whether or not some-one is trying to spend the same coin twice at the same time. The need to have a central register to check each transaction was what forced David Chaum to partner with banks. This necessitated routing all elec-tronic transactions through credit card companies, and re-introduced

5. 'A Cypherpunk's Manifesto', Eric Hughes, 9 March 1993
6. *The Crypto Anarchist Manifesto*, Tim May, 22 November 1992

the Cypherpunks' worst enemies: loss of privacy and the need to trust some hierarchical organisation, a government, bank or corporation, with the authority to verify and, if necessary, roll back transactions.

The solution to the double-spending problem appeared quite suddenly in October 2008, with the publication of a paper on the *The Cryptography Mailing List* entitled *Bitcoin: A Peer-to-Peer Electronic Cash System*. Citing several forerunners in the field, the author of the paper, the previously unknown Satoshi Nakamoto, proposed one key innovation which solved the double-spending problem while preserving anonymity and preventing the need for trusting third parties. This was called the 'blockchain': a distributed ledger, or record of transactions, which would be maintained by everyone participating in the system. It's called the blockchain because groups of transactions are gathered together into 'blocks' as they occur, and as each block is turned out it is added to the 'chain' of all transactions. That's it. It's simply a list of things that happened.

If everyone can see every transaction, then there is no need to hand over control to banks or governments, and if everyone follows the encryption practices of the Cypherpunks, there is no way to know who is spending the money.

Of course, if everyone has a copy of this ledger, we need to know it hasn't been forged or tampered with in any way. So in order to extend the blockchain, in other words to write in the ledger, a certain amount of computational 'work' has to be done: the computer doing the writing has to solve a particularly complex mathematical problem. The fact that it's relatively easy for everyone else's computers to check if this problem really was solved makes it very difficult – in fact, practically impossible – for anyone to create a fake version of the ledger. In a particularly clever twist, participants are incentivised to help maintain the ledger by receiving a small amount of bitcoins when they do solve the mathematical problem. This is where the notional value of Bitcoin comes from: someone has to put in an amount of time and energy to produce it, which is why this process is known as 'mining'. Over time, more and more coins are produced, to an eventual total of 21 million some time in or around 2140.

Satoshi's paper had the good fortune to appear at a particular time. Encoded into the very first block on the Bitcoin chain is a time-stamp, the kind of timestamp more familiar from ransom demands: a proof of life. The phrase embedded forever into the beginning of the blockchain is 'The Times 03/Jan/2009 Chancellor on brink of second bailout for banks', a reference to the front page headline of *The Times* newspaper on that date. On one level, it's a simple proof that no valid coins were mined before that date. On another, it's an ironic comment on the state of the standard economic system that Bitcoin set out to replace. It's also, for those fascinated by such things, one of the earliest clues to the identity of Satoshi Nakamoto.

Satoshi Nakamoto appeared in the world, as far as anyone is aware, with the publication of the Bitcoin white paper. There is no trace of the name before that date, and after a few months of interacting with other developers on the project, Satoshi Nakamoto disappeared just as abruptly from public view at the end of 2010. With the exception of a couple of private emails (indicating that the developer had 'moved onto other things'), and a forum post disavowing an attempt to 'out' the developer in 2014, Satoshi Nakamoto has not been heard from since.

Perhaps instead, more accurately, we might say that the entity referring to itself as Satoshi Nakamoto has not been heard from since 2010. For less interesting than the 'real' identity of Satoshi is the way in which that identity operates in the world – in a way that perfectly accords with Cypherpunk and blockchain doctrine.

In Section 10 of the Bitcoin white paper, Satoshi outlines the privacy model of the system. In the traditional banking model, the flow of money through an exchange is anonymised by the third party administering the transactions; they hide what they know from everyone else. However, on the blockchain, where all transactions are public, the anonymity happens between the identity and the transaction; everyone can see the money moving, but nobody knows whose money it is.

The common idea of cryptocurrencies is that they set assets free, but a cryptocurrency is a monetary unit like in any other currency system – one that, because of the blockchain, is closely monitored and controlled. What's really liberated is identity. It is liberated from responsibility for the transaction and liberated from the 'real' person or persons performing it. Identity, in fact, becomes an asset itself. This is also what marks out the idea of the blockchain from earlier cryptosystems like PGP; it's not the messages that are being hidden, but the actors behind them.

A necessary part of software development is the use of the technology in real-life situations for the purposes of testing. This is often done by the developers themselves in a process known as 'eating your own dogfood.' While the developers of Bitcoin could test mining and transacting coins between them, the real 'product' of Bitcoin – a decentralised, deniable identity – could only be tested by someone (or a group) willing to build and sustain such an identity asset over a long period of time – and who better to perform that test than the creator of Bitcoin themselves? Satoshi Nakomoto is an exercise in dogfooding – and proof of its efficacy.

When Satoshi disappeared into the ether, they left on the blockchain, unspent, the piles of bitcoins they'd personally mined in the early days of the project – over a million of them. These bitcoins are still there, and only someone who holds Satoshi's private keys can access them. Today, Satoshi 'exists' only to the extent someone can prove to be that individual – the only proof of which is possession of those private keys. There is no 'real' Satoshi. There is only a set of assets and a key. 'Satoshi Nakamoto' is creator, product and proof of Bitcoin, all wrapped up in one. Once again, the creation of money is the creation of a myth.

In his book *Debt: The First 5,000 Years*, the anthropologist David Graeber proposes that the connection between finance and sacrifice runs deep in Western culture: 'Why, for instance, do we refer to Christ as the 'redeemer'? The primary meaning of 'redemption' is to buy something back, or to recover something that had been given up in security for a loan; to acquire something by paying off a debt. It is rather striking to think that the very core of the Christian message, salvation itself, the sacrifice of God's own son to rescue humanity from eternal damnation, should be framed in the language of a financial transaction.'[7] Satoshi's sacrifice is something different, but in the anarchic frame from which the individual emerged, not dissimilar. In order to secure the future of Bitcoin, Satoshi gave up all personal gains from its invention: some 980,000 bitcoins, valued at 4 billion dollars in late 2018. This is a gesture that will continue to inspire many in the Bitcoin community, even if few of them understand or even consider its true meaning.

Back in 1995, another regular Cypherpunk contributor, Nick Szabo, proposed a term for the kind of sacrificial identity deployed so success-fully by Satoshi: a 'nym.' A nym was defined as 'an identifier that links only a small amount of related information about a person, usually that information deemed by the nym holder to be relevant to a particular organisation or community.' Thus the nym is opposed to a true name, which links together all kinds of information about the holder, making them vulnerable to someone who can obtain information that is, in the context of the transaction, irrelevant. Or as Szabo put it: 'As in magick, knowing a true name can confer tremendous power to one's enemies.'[8]

Szabo used as examples of nyms the nicknames people used on electronic bulletin boards and the brand names deployed by corpora-tions. The purpose of the nym, in Szabo's reading, is to aggregate and hold reputation in particular contexts: in online discussions on particular topics, or in a marketplace of niche products. But online handles and brand names are not the same things, and their elision is an early echo of the reductionism which the ideology forming around the blockchain would attempt to perform on everything it touched.

---

7. *Debt: The First 5,000 Years*, Chapter Four, David Graeber (Melville House, 2011)
8. 'Smart Contracts Glossary', Nick Szabo, 1995, http://www.fonhum.uva.nl/rob/Courses/
    Information-InSpeech/CDROM/Literature/LOTwinterschool2006/szabo.best.vwh.net/
    smart_contracts_glossary.html

Brand names are a particular kind of untrue name, one associated not merely with reputation but also with financial value. If the brand attracts the wrong kind of attention, its reputation goes down, and so does its value – at least in theory. But because of their value (financial, not reputational), brands also bestow power on the corporations that own them – that know their real name – while often hiding behind them. Brands can sue. They can bribe. They can have activists harassed and killed. Because of their value, brands become things worth maintaining and worth defending. Their goal becomes one of survival, and they warp the world around them to that end.

Online handles are a different kind of untrue name. Their value lies precisely in the fact that they are not tied to assets, not associated with convertible value. They exist only as reputation, which has its own power, but a very different kind. They can be picked up and put down at any time without cost. The key attribute of online handles is not that they render one free through rendering one anonymous, but that they render one free through the possibility of change.

It is precisely this distinction, between financial freedoms and individual autonomy, that underlies many of the debates that have emerged around Bitcoin in recent years, as it struggles to articulate a political vision that is not immured in a technological one. While Bitcoin has proved to be a powerful application for the idea of the blockchain, it has also distorted in the minds of both its practitioners and many observers what the blockchain might actually be capable of.

In many of its practical applications, Bitcoin has so far failed to deliver on its emancipatory promises. For example, one strand of Bitcoin thinking is premised on its accessibility: the widely touted aim of 'banking the unbanked' claims that the technology will give access to financial services to the full half of the world who are currently excluded from real market participation. And yet the reality of Bitcoin's implementation, both technological and socio-political, makes this claim hard to justify. To use the currency effectively still requires a level of technological proficiency and autonomy, specifically network access and expensive hardware, which put up as many barriers to access as the traditional banking system. Regulatory institutions in the form of existing financial institutions, national governments and transnational laws regarding money-laundering and taxation form another barrier to adoption, meaning that to use Bitcoin is either to step far outside the law, into the wild west of narcotics, credit card fraud and the oft-fabled assassination markets, or to participate legally, handing over one's actual ID to brokers and thus linking oneself to transactions in a way that undermines the entire point of an anonymous, cryptographically secure system.

Even if Bitcoin can't emancipate everyone, it might at least do less harm than current systems. Yet in the last couple of years, Bitcoin has made as many headlines for its environmental impact as for its political power. The value of Bitcoin supposedly comes from the computational work required to mine it, but it might more accurately be said that it derives from a more traditional type of mining: the vast consumption and combustion of cheap Chinese coal. It's become terrifyingly clear that the 'mining' of Bitcoin is an inescapably wasteful process. Vast amounts of computational energy, consuming vast quantities of electricity, and outgassing vast quantities of heat and carbon dioxide, are devoted to solving complex equations in return for money. The total power consumption of the network exceeds that of a small country – 42TWh in 2016, equivalent to a million transatlantic flights – and continues to grow.[9] As the value of Bitcoin rises, mining becomes more and more profitable, and the incentive to consume ever more energy increases. This, too, is surely in opposition to any claim to belong to the future, even if one is to take into account the utter devastation imposed upon the earth by our current systems of government and finance.

These complaints, which are both uncomfortably true in the present and addressable in time by adjustments to the underlying system,

---

9. 'Bitcoin's energy usage is huge – we can't afford to ignore it', *The Guardian*, 17 January 2018, https://www.theguardian.com/technology/2018/jan/17/bitcoin-electricity-usage-huge-climate-cryptocurrency

mask the larger unsolved problem posed by the blockchain: *what is it really for?* Somewhere between the establishment of the Cypherpunk mailing list and the unveiling of the first Bitcoin exchange, a strange shift, even a forgetting, occurred in the development of the technology. What had started out as a wild experiment in autonomous self-government became an exercise in wealth creation for a small coterie of tech-savvy enthusiasts and those insightful early adopters willing to take a risk on an entirely untested new technology. While Bitcoin is largely to blame for this, by putting all of the potential of a truly distributed, anonymous network in the service of the market, to focus purely on this aspect of its unfolding is to ignore the potential that remains latent in Satoshi's invention and example. It is to ignore the opportunity, rare in our time, to transform something conceived as a weapon into its opposite.

The arguments over the use of wartime weapons in a time of relative peace, made explicit in the Crypto Wars, have a clear analogue: nuclear technology. While the Allies' desire for global dominance through atomic power was scuppered by Soviet espionage before it began, and the world settled into a Cold War backed by the horrifying possibility of mutually assured destruction, the nuclear powers agreed on one thing: if ever these weapons were to fall into the hands of non-state actors, the results would destroy not merely the social order, but life itself. Similar arguments were made, at the end of the twentieth century, about certain algorithms: the wide availability of cryptography would render toothless the apparatuses of state security and lead to the collapse of ordered society.

While it's easy to scoff now at the idea that the availability of certain complex mathematical processes would bring down governments, we are nevertheless faced with a different, more insidious, threat in the present: that of the substitution of one form of oppressive government with another. While Tim May, part of the original Cypherpunk triumvirate, attested in *The Crypto Anarchist Manifesto* that assassination and extortion markets were 'abhorrent', he had little time for those who weren't part of the crypto utopia. In the sprawling *Cyphernomicon*, a wider exploration of crypto anarchy posted to the Cypherpunks mailing list, May was far clearer on the world he foresaw: 'Crypto anarchy means prosperity for those who can grab it, those competent enough to have something of value to offer for sale; the clueless 95% will suffer, but that is only just. With crypto anarchy we can painlessly, without initiation of aggression, dispose of the nonproductive, the halt and the lame.'[10] Make no mistake: the possibility of cryptologically-enforced fascism is very real indeed. A future where every transaction, financial or social, public or private, is irrevocably encoded in a public ledger which is utterly transparent to those in power is the very opposite of a democratic, egalitarian crypto utopia. Rather, it is the reinstatement of the divine right of kings, transposed to an elevated elite class where those with the money, whether they be state actors, central bankers, winner-takes-all libertarians or property-absolutist anarcho-capitalists, have total power over those who do not.

And yet, as in the nuclear age, there remains space for other imaginaries. In the sixties, in the name of the 'friendly atom', the United States instituted a series of test programs to ascertain whether the awesome power of the atomic bomb could be turned to peaceful ends. Their proposals, some of which were actually carried out, included the excavation of vast reservoirs for drinking water, the exploitation of shale gas (an extreme form of contemporary fracking) and the construction of new roadways. Another idea involved interstellar travel, using the intermittent displacement of atomic bombs in the trail of spacecraft to propel them to distant stars. The former programme was given the name Project Plowshare, in reference to the Prophet Isaiah's injunction to beat swords into plowshares. Long after the cancellation of the project in the face of keen public opposition, the name was taken up by the Plowshares movement, an anti-nuclear weapons and Christian pacifist

---

10. 'Cyphernomicon', Tim May, 1994, https://web.archive.org/web/20180101041641/http://www.cypherpunks.to/faq/cyphernomicron/cyphernomicon.html

organisation that became well-known for direct action against nuclear facilities. Meanwhile, 'peaceful' nuclear energy became a mainstay of everyday life, in the form of the greenest, if most deeply controversial, large-scale energy generation technology we possess. Its outputs, in the form of toxic, radioactive waste, became in turn a source of new contestations over the roles and responsibilities we have to one another, and to the environment.

There is no separation of our technology from the world. Bitcoin, in the decade since Satoshi Nakamoto first announced it, has succeeded technologically but failed politically, because we have failed to understand a central tenet, long established in political theory, that free markets do not create free people – only, and only occasionally, the other way round. A technology developed according to the founding principles of true anarchism – No Gods, No Masters – has already been suborned by capital, because of a lack of imagination and education, and a failure to organise ourselves in the service of true liberation, rather than personal enrichment. This is not a problem of technology, or technological understanding, but of politics.

Bitcoin's touted environmental offences are not a rogue emergent effect, nor the hubristic yet predictable outcome of techno-utopianism. Rather, they are a result of failing to grapple with the central problem of human relations, long diagnosed but rarely put to the test in such dramatic fashion: how to work together in the light of radical equality without falling back into the domination of the rich over the poor, the strong over the weak. But the emergence of that particular offence at this particular time should chime with our position in history. The problem of taking effective global action in leaderless networks is not a problem confined to Bitcoin; in the face of global climate change, it is the primary problem facing humanity today. Like language, the printed word, steam, nuclear power and the internet, another miraculous saviour technology is revealed to be a timely question asked directly to our capacity for change.

At the time of writing, and despite the best, the worst, the most unconsidered and the most deliberate intentions of its progenitors, the blockchain is primarily being used to drive the creation of a new class of monopolists, to securitise existing asset structures, to produce carbon dioxide and to set in stone a regime of surveillance and control unprecedented in the dreams of autocrats. And yet, and yet.

The problem created by blockchain, and dramatised by Bitcoin, is fundamentally inseparable from the political situation it emerged from: the eternal battle between power structures and individual rights. The solution to this problem is not to be found in the technology alone, but in radically different political imaginaries. A word often heard in the corridors of the new blockchain industry seems to encapsulate the inherent contradictions of a cryptologically ordered future; that word is 'trustless.' The concept of trustlessness is at the heart of a vision which seeks to escape from established systems of power by making each individual sovereign to themselves, cryptographically secured, anonymous, untraceable and thus ungovernable. Yet lack of government is but one plank in the construction of freedom: commonality, community and mutual support are equally, if not more, important. This is demonstrated, ultimately, even in the market: as David Graeber has put it, 'the value of a unit of currency is not the measure of the value of an object, but the measure of one's trust in other human beings'.[11] Blockchain, whatever products it might engender in the short term, poses a necessary problem that we should seek to answer not through technological fixes and traditional political forms but through the participation of the widest and most diverse public possible, and the creation of new forms of political relationships between one another.

11. *Debt*, Chapter Three

The White Paper
by Satoshi Nakamoto

Bitcoin: A Peer-to-Peer Electronic Cash System

Satoshi Nakamoto
satoshin@gmx.com
www.bitcoin.org

Abstract. A purely peer-to-peer version of electronic cash would allow online payments to be sent directly from one party to another without going through a financial institution. Digital signatures provide part of the solution, but the main benefits are lost if a trusted third party is still required to prevent double-spending. We propose a solution to the double-spending problem using a peer-to-peer network. The network timestamps transactions by hashing them into an ongoing chain of hash-based proof-of-work, forming a record that cannot be changed without redoing the proof-of-work. The longest chain not only serves as proof of the sequence of events witnessed, but proof that it came from the largest pool of CPU power. As long as a majority of CPU power is controlled by nodes that are not cooperating to attack the network, they'll generate the longest chain and outpace attackers. The network itself requires minimal structure. Messages are broadcast on a best effort basis, and nodes can leave and rejoin the network at will, accepting the longest proof-of-work chain as proof of what happened while they were gone.

1. Introduction

Commerce on the internet has come to rely almost exclusively on financial institutions serving as trusted third parties to process electronic payments. While the system works well enough for most transactions, it still suffers from the inherent weaknesses of the trust-based model. Completely non-reversible transactions are not really possible, since financial institutions cannot avoid mediating disputes. The cost of mediation increases transaction costs, limiting the minimum practical transaction size and cutting off the possibility for small casual transactions, and there is a broader cost in the loss of ability to make non-reversible payments for non-reversible services. With the possibility of reversal, the need for trust spreads. Merchants must be wary of their customers, hassling them for more information than they would otherwise need. A certain percentage of fraud is accepted as unavoidable. These costs and payment uncertainties can be avoided in person by using physical currency, but no mechanism exists to make payments over a communications channel without a trusted party.

What is needed is an electronic payment system based on cryptographic proof instead of trust, allowing any two willing parties to transact directly with each other without the need for a trusted third party. Transactions that are computationally impractical to reverse would protect sellers from fraud, and routine escrow mechanisms could easily be implemented to protect buyers. In this paper, we propose a solution to the double-spending problem using a peer-to-peer distributed timestamp server to generate computational proof of the chronological order of transactions. The system is secure as long as honest nodes collectively control more CPU power than any cooperating group of attacker nodes.

2. Transactions

We define an electronic coin as a chain of digital signatures. Each owner transfers the coin to the next by digitally signing a hash of the previous transaction and the public key of the next owner and adding these to the end of the coin. A payee can verify the signatures to verify the chain of ownership.

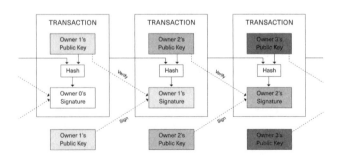

The problem of course is the payee can't verify that one of the owners did not double-spend the coin. A common solution is to introduce a trusted central authority, or mint, that checks every transaction for double spending. After each transaction, the coin must be returned to the mint to issue a new coin, and only coins issued directly from the mint are trusted not to be double-spent. The problem with this solution is that the fate of the entire money system depends on the company running the mint, with every transaction having to go through them, just like a bank.

We need a way for the payee to know that the previous owners did not sign any earlier transactions. For our purposes, the earliest transaction is the one that counts, so we don't care about later attempts to double-spend. The only way to confirm the absence of a transaction is to be aware of all transactions. In the mint-based model, the mint was aware of all transactions and decided which arrived first. To accomplish this without a trusted party, transactions must be publicly announced[1], and we need a system for participants to agree on a single history of the order in which they were received. The payee needs proof that at the time of each transaction, the majority of nodes agreed it was the first received.

3. Timestamp Server

The solution we propose begins with a timestamp server. A timestamp server works by taking a hash of a block of items to be timestamped and widely publishing the hash, such as in a newspaper or Usenet post[2-5]. The timestamp proves that the data must have existed at the time, obviously, in order to get into the hash. Each timestamp includes the previous timestamp in its hash, forming a chain, with each additional timestamp reinforcing the ones before it.

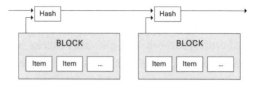

4. Proof-of-Work

To implement a distributed timestamp server on a peer-to-peer basis, we will need to use a proof-of-work system similar to Adam Back's Hashcash[6], rather than newspaper or Usenet posts. The proof-of-work involves scanning for a value that when hashed, such as with SHA-256, the hash begins with a number of zero bits. The average work required is exponential in the number of zero bits required and can be verified by executing a single hash.

For our timestamp network, we implement the proof-of-work by incrementing a nonce in the block until a value is found that gives the block's hash the required zero bits. Once the CPU effort has been expended to make it satisfy the proof-of-work, the block cannot be changed without redoing the work. As later blocks are chained after it, the work to change the block would include redoing all the blocks after it.

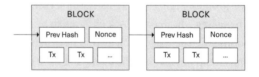

The proof-of-work also solves the problem of determining representation in majority decision making. If the majority were based on one-IP-address-one-vote, it could be subverted by anyone able to allocate many IPs. Proof-of-work is essentially one-CPU-one-vote. The majority decision is represented by the longest chain, which has the greatest proof-of-work effort invested in it. If a majority of CPU power is controlled by honest nodes, the honest chain will grow the fastest and outpace any competing chains. To modify a past block, an attacker would have to redo the proof-of-work of the block and all blocks after it and then catch up with and surpass the work of the honest nodes. We will show later that the probability of a slower attacker catching up diminishes exponentially as subsequent blocks are added.

To compensate for increasing hardware speed and varying interest in running nodes over time, the proof-of-work difficulty is determined by a moving average targeting an average number of blocks per hour. If they're generated too fast, the difficulty increases.

5. Network

The steps to run the network are as follows:

1. New transactions are broadcast to all nodes.
2. Each node collects new transactions into a block.
3. Each node works on finding a difficult proof-of-work for its block.
4. When a node finds a proof-of-work, it broadcasts the block to all nodes.
5. Nodes accept the block only if all transactions in it are valid and not already spent.
6. Nodes express their acceptance of the block by working on creating the next block in the chain, using the hash of the accepted block as the previous hash.

Nodes always consider the longest chain to be the correct one and will keep working on extending it. If two nodes broadcast different versions of the next block simultaneously, some nodes may receive one or the other first. In that case, they work on the first one they received, but save the other branch in case it becomes longer. The tie will be broken when the next proof-of-work is found and one branch becomes longer; the nodes that were working on the other branch will then switch to the longer one.

New transaction broadcasts do not necessarily need to reach all nodes. As long as they reach many nodes, they will get into a block before long. Block broadcasts are also tolerant of dropped messages. If a node does not receive a block, it will request it when it receives the next block and realises it missed one.

6. Incentive

By convention, the first transaction in a block is a special transaction that starts a new coin owned by the creator of the block. This adds an incentive for nodes to support the network, and provides a way to initially distribute coins into circulation, since there is no central authority to issue them. The steady addition of a constant amount of new coins is analogous to gold miners expending resources to add gold to circulation. In our case, it is CPU time and electricity that is expended.

The incentive can also be funded with transaction fees. If the output value of a transaction is less than its input value, the difference is a transaction fee that is added to the incentive value of the block containing the transaction. Once a predetermined number of coins have entered circulation, the incentive can transition entirely to transaction fees and be completely inflation-free.

The incentive may help encourage nodes to stay honest. If a greedy attacker is able to assemble more CPU power than all the honest nodes, he would have to choose between using it to defraud people by stealing back his payments or using it to generate new coins. He ought to find it more profitable to play by the rules, such rules that favour him with more new coins than everyone else combined, than to undermine the system and the validity of his own wealth.

7. Reclaiming Disk Space

Once the latest transaction in a coin is buried under enough blocks, the spent transactions before it can be discarded to save disk space. To facilitate this without breaking the block's hash, transactions are hashed in a Merkle Tree [7,2,5], with only the root included in the block's hash. Old blocks can then be compacted by stubbing off branches of the tree. The interior hashes do not need to be stored.

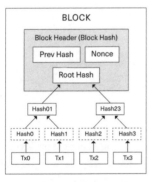

Transactions Hashed in a
Merkle Tree

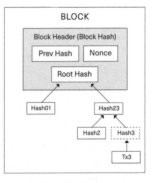

After pruning Tx0-2 from
the Block

A block header with no transactions would be about 80 bytes. If we suppose blocks are generated every 10 minutes, 80 bytes * 6 * 24 * 365 = 4.2MB per year. With computer systems typically selling with 2GB of RAM as of 2008, and Moore's Law predicting current growth of 1.2GB per year, storage should not be a problem even if the block headers must be kept in memory.

8. Simplified Payment Verification

It is possible to verify payments without running a full network node. A user only needs to keep a copy of the block headers of the longest proof-of-work chain, which he can get by querying network nodes until he's convinced he has the longest chain, and obtain the Merkle branch linking the transaction to the block it's timestamped in. He can't check the transaction for himself, but by linking it to a place in the chain, he can see that a network node has accepted it, and blocks added after it further confirms the network has accepted it.

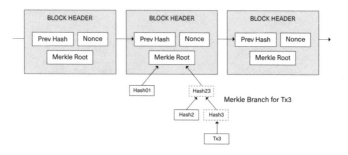

As such, the verification is reliable as long as honest nodes control the network, but is more vulnerable if the network is overpowered by an attacker. While network nodes can verify transactions for themselves, the simplified method can be fooled by an attacker's fabricated transactions for as long as the attacker can continue to overpower the network. One strategy to protect against this would be to accept alerts from network nodes when they detect an invalid block, prompting the user's software to download the full block and alerted transactions to confirm the inconsistency. Businesses that receive frequent payments will probably still want to run their own nodes for more independent security and quicker verification.

9. Combining and Splitting Value

Although it would be possible to handle coins individually, it would be unwieldy to make a separate transaction for every cent in a transfer. To allow value to be split and combined, transactions contain multiple inputs and outputs. Normally there will be either a single input from a larger previous transaction or multiple inputs combining smaller amounts, and at most two outputs: one for the payment, and one returning the change, if any, back to the sender.

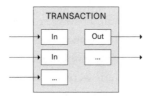

It should be noted that fan-out, where a transaction depends on several transactions and those transactions depend on many more, is not a problem here. There is never the need to extract a complete standalone copy of a transaction's history.

10. Privacy

The traditional banking model achieves a level of privacy by limiting access to information to the parties involved and the trusted third party. The necessity to announce all transactions publicly precludes this method, but privacy can still be maintained by breaking the flow of information in another place: by keeping public keys anonymous. The public can see that someone is sending an amount to someone else, but without information linking the transaction to anyone. This is similar to the level of information released by stock exchanges, where the time and size of individual trades, the 'tape', is made public, but without telling who the parties were.

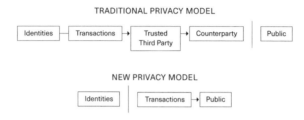

As an additional firewall, a new key pair should be used for each transaction to keep them from being linked to a common owner. Some linking is still unavoidable with multi-input transactions, which necessarily reveal that their inputs were owned by the same owner. The risk is that if the owner of a key is revealed, linking could reveal other transactions that belonged to the same owner.

11. Calculations

We consider the scenario of an attacker trying to generate an alternate chain faster than the honest chain. Even if this is accomplished, it does not throw the system open to arbitrary changes, such as creating value out of thin air or taking money that never belonged to the attacker. Nodes are not going to accept an invalid transaction as payment, and honest nodes will never accept a block containing them. An attacker can only try to change one of his own transactions to take back money he recently spent.

The race between the honest chain and an attacker chain can be characterised as a Binomial Random Walk. The success event is the honest chain being extended by one block, increasing its lead by +1, and the failure event is the attacker's chain being extended by one block, reducing the gap by -1.

The probability of an attacker catching up from a given deficit is analogous to a Gambler's Ruin problem. Suppose a gambler with unlimited credit starts at a deficit and plays potentially an infinite number of trials to try to reach breakeven. We can calculate the probability he ever reaches breakeven, or that an attacker ever catches up with the honest chain, as follows[8]:

p = probability an honest node finds the next block
q = probability the attacker finds the next block
q_z = probability the attacker will ever catch up from z blocks behind

$$q_z = \left\{ \begin{array}{ll} 1 & \textit{if } p \leq q \\ (q/p)^z & \textit{if } p > q \end{array} \right\}$$

Given our assumption that $p > q$, the probability drops exponentially as the number of blocks the attacker has to catch up with increases. With the odds against him, if he doesn't make a lucky lunge forward early on, his chances become vanishingly small as he falls further behind.

We now consider how long the recipient of a new transaction needs to wait before being sufficiently certain the sender can't change the transaction. We assume the sender is an attacker who wants to make the recipient believe he paid him for a while, then switch it to pay back to himself after some time has passed. The receiver will be alerted when that happens, but the sender hopes it will be too late.

The receiver generates a new key pair and gives the public key to the sender shortly before signing. This prevents the sender from preparing a chain of blocks ahead of time by working on it continuously until he is lucky enough to get far enough ahead, then executing the transaction at that moment. Once the transaction is sent, the dishonest sender starts working in secret on a parallel chain containing an alternate version of his transaction.

The recipient waits until the transaction has been added to a block and z blocks have been linked after it. He doesn't know the exact amount of progress the attacker has made, but assuming the honest blocks took the average expected time per block, the attacker's potential progress will be a Poisson distribution with expected value:

$$\lambda = z \; \frac{q}{p}$$

To get the probability the attacker could still catch up now, we multiply the Poisson density for each amount of progress he could have made by the probability he could catch up from that point:

$$\sum_{k=0}^{\infty} \frac{\lambda^k e^{-\lambda}}{k!} \cdot \left\{ \begin{array}{ll} (q/p)^{(z-k)} & if \; k \leq z \\ 1 & if \; k > z \end{array} \right\}$$

Rearranging to avoid summing the infinite tail of the distribution...

$$1 - \sum_{k=0}^{z} \frac{\lambda^k e^{-\lambda}}{k!} \left(1 - (q/p)^{(z-k)} \right)$$

Converting to C code...

```c
#include
double AttackerSuccessProbability(double q, int z)
{
double p = 1.0 - q;
double lambda = z * (q / p);
double sum = 1.0;
int i, k;
for (k = 0; k <= z; k++)
{
    double poisson = exp(-lambda);
    for (i = 1; i <= k; i++)
        poisson *= lambda / i;
    sum -= poisson * (1 - pow(q / p, z - k));
}
return sum;

}
```

Running some results, we can see the probability drop off exponentially with z.

```
q=0.1
z=0    P=1.0000000
z=1    P=0.2045873
z=2    P=0.0509779
z=3    P=0.0131722
z=4    P=0.0034552
z=5    P=0.0009137
z=6    P=0.0002428
z=7    P=0.0000647
z=8    P=0.0000173
z=9    P=0.0000046
z=10   P=0.0000012

q=0.3
z=0    P=1.0000000
z=5    P=0.1773523
z=10   P=0.0416605
z=15   P=0.0101008
z=20   P=0.0024804
z=25   P=0.0006132
z=30   P=0.0001522
z=35   P=0.0000379
z=40   P=0.0000095
z=45   P=0.0000024
z=50   P=0.0000006
```

Solving for P less than 0.1%...

```
P < 0.001
q=0.10    z=5
q=0.15    z=8
q=0.20    z=11
q=0.25    z=15
q=0.30    z=24
q=0.35    z=41
q=0.40    z=89
q=0.45    z=340
```

12. Conclusion

We have proposed a system for electronic transactions without rely-ing on trust. We started with the usual framework of coins made from digital signatures, which provides strong control of ownership, but is incomplete without a way to prevent double-spending. To solve this, we proposed a peer-to-peer network using proof-of-work to record a public history of transactions that quickly becomes computationally impractical for an attacker to change if honest nodes control a major-ity of CPU power. The network is robust in its unstructured simplicity. Nodes work all at once with little coordination. They do not need to be identified, since messages are not routed to any particular place and only need to be delivered on a best effort basis. Nodes can leave and rejoin the network at will, accepting the proof-of-work chain as proof of what happened while they were gone. They vote with their CPU power, expressing their acceptance of valid blocks by working on extending them and rejecting invalid blocks by refusing to work on them. Any needed rules and incentives can be enforced with this consensus mechanism.

References

[1.] W. Dai, 'b-money,' http://www.weidai.com/bmoney.txt, 1998.

[2.] H. Massias, X.S. Avila, and J.-J. Quisquater, 'Design of a secure timestamping service with minimal trust requirements,' in *20th Symposium on Information Theory in the Benelux*, May 1999.

[3.] S. Haber, W.S. Stornetta, 'How to time-stamp a digital document,' in *Journal of Cryptology*, vol 3, no 2, pages 99-111, 1991.

[4.] D. Bayer, S. Haber, W.S. Stornetta, 'Improving the efficiency and reliability of digital time-stamping,' in *Sequences II: Methods in Communication, Security and Computer Science*, pages 329-334, 1993.

[5.] S. Haber, W.S. Stornetta, 'Secure names for bit-strings,' In *Proceedings of the 4th ACM Conference on Computer and Communications Security*, pages 28-35, April 1997.

[6.] A. Back, 'Hashcash - a denial of service counter-measure,' http://www.hashcash.org/papers/hashcash.pdf, 2002.

[7.] R.C. Merkle, "Protocols for public key cryptosystems," In Proc. 1980 *Symposium on Security and Privacy*, IEEE Computer Society, pages 122-133, April 1980.

[8.] W. Feller, "An introduction to probability theory and its applications," 1957.

The White Paper
by Satoshi Nakamoto

Guide by Jaya Klara Brekke

Bitcoin: A Peer-to-Peer Electronic Cash System

Satoshi Nakamoto

satoshin@gmx.com

www.bitcoin.org

Abstract. A purely peer-to-peer version of electronic cash would allow online payments to be sent directly from one party to another without going through a financial institution. Digital signatures provide part of the solution, but the main benefits are lost if a trusted third party is still required to prevent double-spending. We propose a solution to the double-spending problem using a peer-to-peer network. The network timestamps transactions by hashing them into an ongoing chain of hash-based proof-of-work, forming a record that cannot be changed without redoing the proof-of-work. The longest chain not only serves as proof of the sequence of events witnessed, but proof that it came from the largest pool of CPU power. As long as a majority of CPU power is controlled by nodes that are not cooperating to attack the network, they'll generate the longest chain and outpace attackers. The network itself requires minimal structure. Messages are broadcast on a best effort basis, and nodes can leave and rejoin the network at will, accepting the longest proof-of-work chain as proof of what happened while they were gone.

There is nothing like the pleasure of a perfect mental model, a puzzle solved. For the technically-minded, the Bitcoin white paper presents a stunningly elegant solution to a set of hugely complex problems. The unusual combination of decentralised network architecture, cryptography and economic incentives promises a system that can run independently and can also be secure, economically viable and beyond the reach of anyone seeking to control or shut it down. This vision is what makes so many people's eyes shine with excited fervour when recounting their first reading of this paper. Implied in this technical architecture is a proposal for a peer-to-peer money system that would remove the need for authorities and thereby circumvent banks, governments and legal systems.

This particular formula and abstract template, through blockchain, has since been expanded from electronic payments to any type of data or computation; a Turing-complete machine for arranging, registering and enforcing shared truths about events. The unusual simplicity of the solution has become so convincing that the scandals, contradictions, manipulation and messiness of actual implementations are often over-looked as irrelevant noise from a messy and indeterminate world. The white paper, in contrast, projects a promise of clean determinacy, a functional truth founded on mathematically-arranged decentralisation.

The first key to comprehending this system is to understand decentralisation and where this operationalisation of the word comes from. Bitcoin is one of the more famous descendants in a lineage of decentralised technologies developed explicitly to defeat authority. It is one step further from distributed systems, which are resilient to fail-ures, faults and accidents by making sure that information is stored in many different places and can take many different routes to where it needs to get to. Decentralisation goes beyond this, aiming to be resilient to censorship, manipulation and shutdown. A decentralised architecture does this by having several or no 'authorities', such that no single aspect of the system is fully relied upon.

This idea of decentralisation came out of a set of specific expe-riences from the late eighties to the early noughties. The internet was an open network of free flows of information, ideas, films, music and knowledge, but then individuals began to be arrested, servers seized and sites shut down for breach of copyright and intellectual property. In response, decentralised systems were set up, such that even if a particular person or server was taken down by authorities, the systems themselves (the files, music, films, data and so on) would still be avail-able. The relay system Tor was established to allow for anonymous web browsing, while BitTorrent was established to enable the direct and untraceable sharing of files in response to the shutdown of music shar-ing platform Napster. The concerns that led to the setting up of these decentralised systems also extended to payments and money systems; if networks were being targeted, monitored and controlled by author-ities, and an increasing amount of trade and transactions were taking place via networks, then these were vulnerable to financial control and censorship. There were several experiments in the decentralisation of money: DigiCash was created specifically to address privacy concerns with regards to electronic payments, while E-Gold was based on the concept of gold as 'superior' money. Both the concern for privacy and the desire for a gold-like currency were carried through into Bitcoin, encoding aspects of a libertarian understanding of money, power and economics but born out of a network strategy to circumvent authority through the use of a decentralised architecture.

In early 2009, the global financial system was in the middle of a catastrophic crash, the British Chancellor was on the 'brink of second bailout for banks' – and the first transaction of a new and decentralised digital payment system, Bitcoin, took place[A]. The Bitcoin logo began to appear amongst the people, tents, dogs and banners that filled the squares and streets of cities across the world as people reacted in anger to betrayal by politicians, financial and legal authorities.

This was the motivation for decentralisation: to create a network beyond the control of authorities that would withstand censorship and potential shutdown by even the most powerful corporate entities or governments. With Bitcoin, this very specific idea of decentralisation shifted from a specific strategy of circumvention to a political proposition in its own right. The strategic tools of decentralised architectures, cryptography and networks became a possible solution to any kind of authority; they became a systems primacy.

With the following notes I want to make your eyes shine in bright-eyed wonder as you re-read the Bitcoin white paper, just as mine did – but I also want to share my tools for picking apart how the machine works. I want to pass on some secret keys with the hope of unlocking a sensibility that might reach beyond bright lights towards important new friendships and alliances. What I mean is this: let's not replace one set of authorities for another, albeit machinic one. The Bitcoin white paper generated an explosion in new ideas for money systems,

A. The very first Bitcoin transaction on 3 January 2009 included a hash of that day's article in *The Times* newspaper https://www.thetimes.co.uk/article/chancellor-alistair-darling-on-brink-of-second-bailout-for-banks-n9l382mn62h

political systems and legal systems too. In the meantime, the system itself turned out to be not so decentralised, not so secure, not so objective, not so deterministic. The computational network might continue running in the face of authorities, but in the meantime, people are arrested, funds are stolen and lost, exchange rates are manipulated, wealth is accumulated, new 'authorities' form in the shadows and new suckers are blinded by the bling of Initial Coin Offerings. *It is not the fault of the system*, you say. *It is the people that have the faults*, I hear. But take note: what is true in a model might not be true for the realities it creates. The system itself might have been made safe from risk and shutdown by authorities, but those using it most certainly have not.

A decentralised computer network does not equal decentralised power. The biggest mistake, the most pitiful temptation, is to conflate the truth of a stunning systems model with the truth of an infinitely more complex, alive and indeterminate universe. Read the white paper. But then take your eyes off this piece of paper – and look up.

1. Introduction

Commerce on the internet has come to rely almost exclusively on financial institutions serving as trusted third parties to process electronic payments. While the system works well enough for most transactions, it still suffers from the inherent weaknesses of the trust-based model.

Trust *I don't trust in politicians, but I trust in maths*; so said a man at a London Bitcoin meetup some years ago. It's a sentiment shared by many in the Bitcoin communities. Who can be trusted? In an open decentralised network, the answer is no one. Any node in the network might be malicious, and if a given system depends on – or *trusts* – this particular node, then the whole system is compromised. This is why a 'trust-based model' is 'inherently weak'. In the design of decentralised network architectures, the aim is to get as close to 'trustlessness' as possible, so that no single aspect of the system is fully relied upon for it to function as intended. 'Trustlessness' therefore also means transparent and open. If there is any piece of code or activity that is not fully visible or known, this implies trust – and this means a security weakness that might be exploited. In the context of the financial and subsequent sovereign debt crises, the promise of a system that would function without the need to trust in any authorities took on a much broader meaning and became an enticing social and political prospect. A system built with the assumption that no person, authority or institution can be trusted implies that only the *trustless* system itself can be trusted. But in the meantime, a lot of trust indeed is required for most people to use Bitcoin; trust in the explainer you read on a website that showed up when you typed 'Bitcoin' into Google, in a particular cryptocurrency exchange where you are buying your bitcoin and the rates they are giving you, in your handling of cryptographic keys, in the device you are using and so on. For whom is the system 'trustless', then? Does a trustless system help build trust amongst communities, or does it spread more trustlessness, more risk, more insecurity? Can a lack of social and political trust and security be resolved through other means than technical trustlessness?

> Completely non-reversible transactions are not really possible, since financial institutions cannot avoid mediating disputes. The cost of mediation increases transaction costs, limiting the minimum practical transaction size and cutting off the possibility for small casual transactions, and there is a broader cost in the loss of ability to make non-reversible payments for non-reversible services.

Mediation Getting rid of middlemen: this has long been the aim of peer-to-peer systems, which understand mediators as, at best, an unnecessary cost and, at worst, rent-seekers and security threats. But what is mediation? When is it an unnecessary cost? When does it, instead, help to build understanding and mitigate risk? Is it possible to get rid of mediation entirely, or are all our experiences to some extent mediated? From a network security standpoint in decentralised systems, mediation presents a weak point, in that the mediator might herself be corrupt or malicious – or might become the target of external authorities, looking to impose their will by targeting a weak point in the system. Financial institutions have the power to reverse transactions and might do so, for example, in the case of economic sanctions. Bitcoin was posited as a disinterested system that would execute exactly as coded, with no recourse to influence. Decentralised networks promise free and unstoppable flows that can circumvent arbitrary geopolitical vagaries or corrupt individuals, beyond the control of authorities – in fact, beyond the control of anybody. But in order to do so, a decentralised protocol itself takes the place of the mediator and becomes a new form of intermediation; one that speaks a different language and has unknown accountability structures. Yet mediation and reversibility are, in other circles, considered positive. This is also the case in payment systems: if something goes wrong, in cases of dispute, there is the possibility of an appeal to some external mediator that can help resolve the dispute. And as it turned out, the system itself was not quite so disinterested or neutral. This became clear in the Bitcoin scaling conflict, which has also been referred to as Bitcoin's constitutional moment.

> With the possibility of reversal, the need for trust spreads. Merchants must be wary of their customers, hassling them for more information than they would otherwise need. A certain percentage of fraud is accepted as unavoidable. These costs and payment uncertainties can be avoided in person by using physical currency, but no mechanism exists to make payments over a communications channel without a trusted party.

Cash The intention of Bitcoin – to become a kind of cash for the internet – was spectacularly justified early on in its history. In 2011, the whistleblower project Wikileaks faced an unofficial banking blockade by US companies facilitating most global payments. Wikileaks had published a leak by US soldier Chelsea Manning documenting US military drone strikes and the killing of thousands of civilians in Iraq and Afghanistan. The resulting banking blockade became, for many, justification for the need for a neutral global payment system that could not be censored and that was not controlled solely by the US government. The benefit of a physical currency, as opposed to digital payments, is that an exchange takes place directly between two people. Digital payments are facilitated by private companies (Paypal, banks, Visa, SWIFT and so on) who then hold all your payment information and can potentially make this available to other companies or authorities. Cash cannot be easily traced; it has an element of anonymity and it is 'fungible', meaning that it is an equaliser in the sense that a pound or a dollar is a pound or a dollar regardless of who carries and tries to spend it. There were strong disagreements in early discussions on the *Bitcointalk* forum about whether Bitcoin should publicly support Wikileaks, and thereby become an explicit strategy for circumventing authority, or aim for everyday use first: 'I say, we MUST get Bitcoin accepted at Starbucks and the local grocery store.... BEFORE it gets accepted at Wikileaks.'[B] Some were worried that political association with the organisation would kill the project from its infancy: 'Currently all intelligence services are working to smash wikileaks. I would not want to suddenly posthumously come into Al-Qaeda.'[C] Others argued that this was the moment to prove the efficacy and necessity of a system like Bitcoin: 'Paypal just blocked them, and they're trying to get other US banks do the same. This would be a great moment to open bitcoin donations' and 'bring it on. Let's encourage Wikileaks to use bitcoins and I'm willing to face any risk or fallout from that act.'[D, E] The inventor/s Satoshi Nakamoto disagreed: 'No, don't 'bring it on, the project needs to grow gradually so the software can be strengthened along the way. I make this appeal to WikiLeaks not to

B.   See Bruce Wagner https://bitcointalk.org/index.php?topic=1735.msg26814#msg26814
C.   See bitcoinex https://bitcointalk.org/index.php?topic=1735.msg25360#msg25360
D.   See Wumpus https://bitcointalk.org/index.php?topic=1735.msg26737#msg26737
E.   See RHorning https://bitcointalk.org/index.php?topic=1735.msg26876#msg26876

try to use Bitcoin. Bitcoin is a small beta community in its infancy. You would not stand to get more than pocket change, and the heat you would bring would likely destroy us at this stage.'[F]

Wikileaks founder Julian Assange apparently discussed the matter with Nakamoto and agreed to delay the public affiliation for another few months. Nevertheless, one of the final comments by Satoshi Nakamoto was this: 'It would have been nice to get this attention in any other context. WikiLeaks has kicked the hornet's nest, and the swarm is headed towards us'.[G] After this comment, Satoshi Nakamoto largely disappeared. Years later, in 2017, the Bitcoin network forked as a result of the Bitcoin scaling conflict, and a new version, Bitcoin Cash, was launched. The scaling conflict was about how best to scale the network, but it was also to some degree about whether Bitcoin should replace existing payments systems and be able to facilitate billions of small transactions, or whether it should be a digital settlement layer, potentially replacing the dollar as a global standard, facilitating cash-like layers on top. The conflict was essentially about the purpose of Bitcoin. In what way exactly should Bitcoin be a disruptive technology? By outcompeting existing payment infrastructures in terms of existing criteria? Or by being a layer that facilitates circumvention of authority? The voice of the absent Satoshi Nakamoto has been conjured to support claims that Bitcoin was 'meant' to be cash, but the open decentralised characteristics of the project do not easily conform to authority of any kind. Disagreements between a vision of mainstream adoption or anti-authoritarian strategy still continue to this day in various forms: The Bitcoin Foundation was established as an attempt to represent Bitcoin amongst legislators, to make it a legitimate and broadly accepted means of payment, but a decentralised network cannot be so easily represented. The dark

F.   See Satoshi https://bitcointalk.org/index.php?topic=1735.msg26999#msg26999
G.   See Satoshi https://bitcointalk.org/index.php?topic=2216.msg29280#msg29280

vision for Bitcoin has remained strong amongst those who understand the project primarily as a strategy against authorities – a means to maintain a darknet where the floodlights of authority cannot reach.

> What is needed is an electronic payment system based on crypto-graphic proof instead of trust, allowing any two willing parties to transact directly with each other without the need for a trusted third party.

Cryptographic Proof Here is the mathematical 'magic' of Bitcoin: the ability to assess the integrity of records and verify new data, instead of relying on the bank, the state, the law or the authority of any other insti-tution. How? Cryptographic hashing is a mathematical phenomenon whereby some data is run through a hashing algorithm, which will then output a string of characters. Only this particular data run through this particular hashing algorithm will produce that specific string of char-acters. What this means is that if the data is tampered with in any way, this can be proven by running it through the hashing algorithm again and checking if the output is the same. This peculiar phenomenon is the basis for a variety of cryptographic applications that are used in the Bitcoin payment system: public key encryption, data verification and decentralised consensus. Instead of trusting in what the bank says, here is cryptographic proof instead. In the hands of a skilled person, cryptography can protect secrets, messages and knowledge from even the most powerful armies and authorities. But when Bitcoin became a generalised political idea instead of just a strategy, crypto-graphy became an ideology instead of just a tool. Replacing 'trust' with cryptographic proof promised to replace the uncertainty of humans and institutions with the certainty of mathematics – and yet cryptographic hashing is itself an area of human endeavour, and indeed an arms race. There are many different hashing algorithms that have different characteristics – Bitcoin uses SHA-256 – and as computers become more powerful, new algorithms must be developed in order to guar-antee security properties (see, for example, the NIST competition). For believers, hashing is a discovery of a universal principle that all else can be anchored to; for others, it is an area of research and invention – an ongoing effort of work that has the potential to tip the balances of power, but is also an area of creativity, knowledge sharing and collaboration amongst peers experimenting and working with the phenomena of this curious universe that we live in.

> Transactions that are computationally impractical to reverse would protect sellers from fraud, and routine escrow mechanisms could easily be implemented to protect buyers. In this paper, we propose a solution to the double-spending problem using a peer-to-peer distributed timestamp server to generate computational proof of the chronological order of transactions.

Double-spending Problem How does a decentralised network reach consensus on which transactions are true? A peer-to-peer payment system comes up against some curious problems. Normally, a bank holds a record of transactions and registers that a transaction has taken place, as well as who the transaction was between, when it occurred and how much money was transferred. In other words, the bank is the authority that takes note of and determines what has taken place and what transaction is true. In Bitcoin, the intention is to get rid of authority and have transactions facilitated by and for the network itself. So, instead, everyone (potentially) holds the record of accounts, and (potentially) verifies transactions (the details of how this takes place will be described later). But if the record of transactions is held by multiple nodes across a network, how does the network agree on which transactions to consider valid? A person might tell one part of the network about a transaction to recipient 'A', and then tell another part of the network of a different recipient 'B'. This is called the double-spending problem. In network computation, this is also called the Byzantine Generals' Problem; how can one know which messages are true when they are passed through potentially untrusted messengers? (See appendix, page 73, for Satoshi

Nakamoto's explanation of the problem). When there is no longer an authority that determines which of these transactions are valid, as in a peer-to-peer system, this must now be done through a mechanism in the decentralised network. The Bitcoin consensus mechanism proposes a means for achieving decentralised consensus at scale. But this raises a set of fundamental questions; what kind of consensus is organised through the Bitcoin protocol? How does computational consensus differ from social or political consensus? And how is dissensus accommodated for?

> The system is secure as long as honest nodes collectively control more CPU power than any cooperating group of attacker nodes.

2. Transactions

We define an electronic coin as a chain of digital signatures.

Coins (Chains) Bitcoins are not 'coins' exactly. Instead, a chain of digital signatures determines how a unit is passed between different 'owners', where the previous owner signs a value over to the next owner by including their *public key* in the transaction (see *Public-key cryptography* below). Bitcoin as a ledger of transactions and bitcoin mining as analogous to gold mining point to two very different theories of money, and the project has therefore attracted a range of people with very different understandings of money, economics, value and the world. On one hand, the blockchain digital ledger is reminiscent of a credit theory of money, specifically money as a series of IOUs. The credit theory of money considers money tokens as representative of relationships of credit and debt, where a coin or a record of accounts are ways of making these relationships visible. A ledger can do this by registering credits and debts, and physical coins also do this by individuals possessing coins (having credit) or not possessing coins (having no credit). If money is a representation of specific kinds of relationships, what kinds of relationships are they? How are these relationships organised, chained together and enforced? How might they be organised differently? These ideas conflict with *commodity theories of money*, also encoded into the Bitcoin architecture through gold analogies, which tend to believe in an inherent value of things and a natural state of money and markets (this will be explained more thoroughly below).

> Each owner transfers the coin to the next by digitally signing a hash of the previous transaction and the public key of the next owner and adding these to the end of the coin.

Public-key Cryptography Public-key cryptography is used in all manner of systems to make sure that a message is only read by its intended recipient. Cryptographic hashing, the mathematical magic described earlier, can be used to create a set of keys: one that can encrypt (your public key) and another that can decrypt (your private, secret key). You can share your public key with the world, so that anyone can write a message and lock (encrypt) it with your public key, meaning that it cannot be read. Only you (or someone who has stolen your private key) can unlock, decrypt and read that message. The public key encrypts messages to be sent to the owner; the owner then uses the private key to decrypt messages.

> A payee can verify the signatures [...]

Digital Signatures Cryptographic keys can also be used as a way to sign a message so that someone knows that it is coming from you and not an impersonator. In Bitcoin, the 'message' is a transaction. You use your private key to sign a transaction that you want to send to someone. The network can then verify that the transaction is indeed coming from you, the 'owner', by checking it against your public key.

> [...] to verify the chain of ownership.

Chains of ownership (from pirates to police) In Bitcoin, public key cryptography and digital signatures are chained together to form a linear history of changing ownership: Owner0 signs a hash of their previous transaction, along with the public key of the new owner1. When owner1 wants to send transaction to owner2, they use their private key to sign a hash of their old transaction with owner0 along with the public key of the new owner2, and so on. Each transaction is signed by the current owner and chained to the previous transaction and the intended recipient's public key – who thereby becomes the new owner. Only the person holding the corresponding private key can access and spend the transaction. With Bitcoin, then, cryptographic keys have taken on an entirely new function: they are a way to define and enforce property. Whereas cryptographic tools were previously used primarily for secrecy and protecting privacy, mathematical keys now 'lock' property and enable the design and enforcement of any manner of access criteria. This also radically alters the economic imagination of much of the peer-to-peer movement in ways that have so far gone almost entirely uncommented on: from an internet culture that was critical of things like intellectual property because of a preference for open, fluid networks and abundance of knowledge, towards a generation of peer-to-peer blockchain developers who are heavily invested in determining, designing and enforcing property and property rights in ever more fine-grained ways through automated algorithmic authorities.

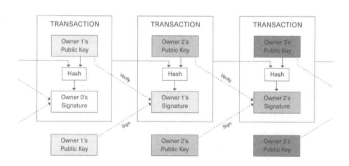

The problem of course is the payee can't verify that one of the owners did not double-spend the coin. A common solution is to introduce a trusted central authority, or mint, that checks every transaction for double spending. After each transaction, the coin must be returned to the mint to issue a new coin, and only coins issued directly from the mint are trusted not to be double-spent. The problem with this solution is that the fate of the entire money system depends on the company running the mint, with every transaction having to go through them, just like a bank.

Public ledger ('unpermissioned') To get rid of authorities is, potentially, the same as making everyone an authority. In order for Bitcoin to be decentralised, it cannot rely on a specific authority to record and verify transactions. Instead, anyone can take part in those tasks. But because the system is open and anyone, including a dishonest person, might contribute to verifying transactions, the system needs to be made secure. To this end, all transactions, and all verified blocks of transactions, are publicly announced, witnessed and recorded with cryptographic proofs that can be checked (see *Double spending* and *Cryptographic proofs* above). In later years, this open system design became known as an 'unpermissioned' blockchain mechanism, because it does not require special permissions for anyone to download and run the client so they can take part witnessing and verifying the network or browsing the records. Such a level of public scrutiny does not serve certain businesses very well – so new blockchain and cryptocurrency systems were developed to require special permissions in order to write or read data on the chain. 'Permissioned' chains mean that new and fine-grained ways of managing the recording and reading of data are made possible. The distinction between permissioned and unpermissioned chains represent a political distinction in terms of how decentralisation is viewed. Decentralisation, from a network design perspective, is only strictly true for unpermissioned systems. A permissioned system implies that those with permissions are in a sense new authorities, and the system would require trust in these. Yet the ways that permissioned or unpermissioned systems relate to decentralisation in terms of power, politics and economics are far more complex. For some communities, a permissioned system might be more secure against authorities than a fully open, unpermissioned system, and although the system might be permissionless, this does not necessarily mean it is accessible. Permissions (managed through public-key cryptography) represent only one kind of management of accessibility. Hardware capacity, technical capacity and other aspects of the system can make a seemingly open system inaccessible for many.

> We need a way for the payee to know that the previous owners did not sign any earlier transactions. For our purposes, the earliest transaction is the one that counts, so we don't care about later attempts to double-spend. The only way to confirm the absence of a transaction is to be aware of all transactions. In the mint-based model, the mint was aware of all transactions and decided which arrived first. To accomplish this without a trusted party, transactions must be publicly announced[1], and we need a system for participants to agree on a single history of the order in which they were received.

Time (the blockchain) A timestamp server is used to determine when a transaction is witnessed in the network. Transactions in a given period of time are then grouped into a block and hashed with the timestamp. The timestamp proves when the transactions were seen, and where they belong in the linear record of transactions. The block includes the timestamp of the previous block so that it too forms a chain – a *blockchain* – of timestamped transactions. The blockchain forms a linear history of transactions from the very first bitcoin. But who gets to hash the transactions into a block? How can we make sure that it is not the same node each time? If it was the same node each time, this node would essentially be an authority, deciding which transactions are valid or not. And what if there are contradicting transactions or double-spending attempts? How does the network agree on which block of transactions to consider valid?

The payee needs proof that at the time of each transaction, the majority of nodes agreed it was the first received.

3. Timestamp Server

The solution we propose begins with a timestamp server. A timestamp server works by taking a hash of a block of items to be timestamped and widely publishing the hash, such as in a newspaper or Usenet post[2-5]. The timestamp proves that the data must have existed at the time, obviously, in order to get into the hash. Each timestamp includes the previous time-stamp in its hash, forming a chain, with each additional timestamp rein-forcing the ones before it.

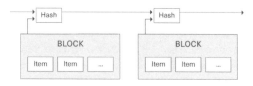

4. Proof-of-Work

To implement a distributed timestamp server on a peer-to-peer basis, we will need to use a proof-of-work system similar to Adam Back's Hashcash[6], rather than newspaper or Usenet posts.

Proof-of-work (mining 1/5) Nodes in the network broadcast and witness transactions. Other nodes gather these into blocks and hash them into a block, referencing a previous block to form a chain of verified transactions: the blockchain. If the same node could always determine the next block, then it would essentially be able to control which transactions are considered valid and would also be able to manipulate the records. It would become, so to speak, a (potentially corrupt) authority. To avoid this, it's necessary to have a way of fairly deciding which node gets to add the next block. The network has to agree on the block and to make sure that it cannot be tampered with. This is done through a competition similar to a lottery, using mathematical probability: computers in the network are set to run a hashing algorithm repeatedly, hashing a block of transaction data along with arbitrary numbers, referred to as a nonce. They try different nonces (random numbers) until they find one that, when hashed with the block of transaction data, spits out a string that meets certain requirements; specifically, it has to have a certain amount of zeros, like this: 000000000019d6689c085ae165831e934ff763ae46a2a6c172b3f-1b60a8ce26f (this hash is from the *genesis block* – the very first block of the Bitcoin blockchain). The hashed output is called the proof-of-work, and is cryptographic proof that a node has done the mining work. It is (probabilistically) impossible to fake the result, and the nonce and transactions are published so that anyone can check if indeed, when run through the hashing algorithm, they correspond to the published hash. This is the cryptographic proof that the computational 'work' was done, thereby validating the block as the next one to be added to the chain. A new round of competition then starts to find the next nonce and to be the node that gets to decide the next block of transactions. This is what 'consensus' means in the particular, peculiar and probability-based security model of Bitcoin: mathematically determined events to decide what is or is not true in the network and should be added to the ledger.

The idea of forcing nodes to do difficult work as a security and access mechanism draws on earlier work in the field of computation. In 1992, computer scientists Cynthia Dwork and Moni Naor suggested using a difficult function as a way to combat junk mail – or indeed any abuse of resources – by making it costly in terms of time and computational effort. The idea was that if some minimal effort was required it would not be worthwhile for junk mailers to spam the system. This idea was taken up by Adam Back, and developed into his Hashcash idea. Published in 1997, it outlines how denial-of-service (DoS) attacks and other kinds of abuse of internet resources could be prevented by requiring a proof-of-work token that would prove that the necessary work had been done. These were the seeds of the curation of scarcity in an otherwise infinitely replicable digital space. This was the beginning of incorporating economic dynamics deep into network protocol design. This is Bitcoin mining, a first round of explanation.

The proof-of-work involves scanning for a value that when hashed, such as with SHA-256, the hash begins with a number of zero bits. The average work required is exponential in the number of zero bits required and can be verified by executing a single hash.

For our timestamp network, we implement the proof-of-work by incrementing a nonce in the block until a value is found that gives the block's hash the required zero bits. Once the CPU effort has been expended to make it satisfy the proof-of-work, the block cannot be changed without redoing the work. As later blocks are chained after it, the work to change the block would include redoing all the blocks after it.

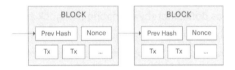

The proof-of-work also solves the problem of determining representation in majority decision making. If the majority were based on one-IP-address-one-vote, it could be subverted by anyone able to allocate many IPs. Proof-of-work is essentially one-CPU-one-vote.

CPU voting (mining 2/5) Computers (in fact, factories full of them) compete to be the next ones to determine and validate a block of transactions. They run the hashing algorithm with transaction data repeatedly, adding different random numbers as quickly as possible to identify one that gives an output with the correct amount of zeros in front. The idea was that Central Processing Units (CPU) would ensure fair and open participation in verifying transactions and mining Bitcoin. Nakamoto's white paper compares this to voting, because determining a block also signals a decision on which chain this block is added to; a 'vote' on which record of transactions should become the accepted history of events. There are times when the chain forks; for instance, if there is disagreement about which Bitcoin transactions are valid or even which Bitcoin client is correct (see below). The accepted history of events, which we can consider to be the consensus in the network, is therefore an emerging property, as more blocks add 'consensus' to a given chain. Yet not all voting efforts have proven to be equal. Very quickly, as the competition to mine blocks intensified, new and dedicated hardware was developed specifically to run the Bitcoin hashing algorithm. From *Central Processing Units to Graphics Processing Units to Application Specific Integrated Circuits*, CPU-> GPU->ASIC-voting, the difficulty of mining increased and moved out of reach for most ordinary computers. The resulting centralisation of Bitcoin mining remains one of its major governance problems, derived from the fact that the system's design is reliant on specific, externally-produced hardware. The high energy cost of that hardware is one of its major environmental problems too. These are areas of intense and politicised development that many subsequent cryptocurrency and blockchain designs address in different ways, including chip and hardware designs and their relationship with modified consensus mechanisms that look to prevent centralising tendencies in the networks.

The majority decision is represented by the longest chain, which has the greatest proof-of-work effort invested in it. If a majority of CPU power is controlled by honest nodes, the honest chain will grow the fastest and outpace any competing chains. To modify a past block, an attacker would have to redo the proof-of-work of the block and all blocks after it and then catch up with and surpass the work of the honest nodes. We will show later that the probability of a slower attacker catching up diminishes exponentially as subsequent blocks are added.

To compensate for increasing hardware speed and varying interest in running nodes over time, the proof-of-work difficulty is determined by a moving average targeting an average number of blocks per hour. If they're generated too fast, the difficulty increases.

5. Network

The steps to run the network are as follows:

1. New transactions are broadcast to all nodes.
2. Each node collects new transactions into a block.
3. Each node works on finding a difficult proof-of-work for its block.
4. When a node finds a proof-of-work, it broadcasts the block to all nodes.
5. Nodes accept the block only if all transactions in it are valid and not already spent.
6. Nodes express their acceptance of the block by working on creating the next block in the chain, using the hash of the accepted block as the previous hash.

A node (is not a node) The Bitcoin network(s) have, over time, developed their own governance systems and balances of power. These network governance methods work quite differently to other forms of institutional governance; a different arrangement of code and people, accountability and privacy, geography and power. Yet they face many of the same problems, not least a concentration of power amongst those with the capacities and sensibilities demanded by this specific system to operate effectively within its territory. Typical decentralised network diagrams show nodes communicating with other nodes of equal sizes that are close to them in what looks like a harmonious and horizontal arrangement. It looks democratic, and it gives an intuitional sense of equality. Actual networks are more complex than that; they might be designed to communicate with nodes according to all kinds of criteria, including closest, farthest, most capacity, best reputation or otherwise. They also change over time and have important emergent characteristics. Here, the white paper describes in the above steps what nodes in the Bitcoin network do. The network is open and the client can be downloaded by anyone wishing to run a Bitcoin node, but in the years since the white paper was published and the first Bitcoin transaction was registered, these tasks have become increasingly more specialised. Not all nodes necessarily '[collect] new transactions into a block'; Bitcoin wallets that will simply broadcast and witness have been developed. Not all nodes 'work on finding a difficult proof-of-work for [their] block[s]'; full nodes that might broadcast and witness transactions and check that these comply with the consensus rules, but not mine blocks, have emerged. Not all nodes 'accept the block'; many don't even look at the blockchain. Mining has become so competitive and difficult that mining pools have developed. These pools look more like a federated system than a decentralised system. And an entirely new layer, a *lightning network*, is developed on top to allow faster transactions. In other words, a node is not just a node, and a network is not just a network. And thus as a living system they continue to evolve into specialised roles and clusters of networks with different topologies: 'decentralised', 'federated', 'supernode', 'centralised', 'striated' and so on.

> Nodes always consider the longest chain to be the correct one and will keep working on extending it. If two nodes broadcast different versions of the next block simultaneously, some nodes may receive one or the other first. In that case, they work on the first one they received, but save the other branch in case it becomes longer. The tie will be broken when the next proof-of-work is found and one branch becomes longer; the nodes that were working on the other branch will then switch to the longer one.

Longest chain (chain forks and scaling conflict) Should difference be mediated by a single protocol or 'fork' into multiple and multiplying protocols? Should these different protocols then be compatible? Are true differences incompatible? Does incompatibility mean exclusion or autonomy? Is freedom the choice between compatibility or incompatibility? It turns out that not everyone always agrees on which transaction is correct, or which history of transactions is correct, or even which version of the Bitcoin protocol is correct. What happens when there is no consensus in a decentralised network: for transactions, if contradictory blocks are mined, these create a fork in the blockchain. Eventually, one fork will 'win', based on the rule that the longest chain is true. In other words, the protocol rules are written so that the network will eventually agree on a single history of events, settling on the fork that over time becomes the longest chain, the one that most miners have mined on; the one that has most computational 'consensus' (of quite a different kind than social consensus). But what if the network protocol itself is not agreed on? Forking is not only a temporary happenstance of a network as it seeks consensus; it has become a *dissensus mechanism* more generally.

Forking is a concept and technique from the world of open-source software development that has, through Bitcoin and over the course of major conflicts in cryptocurrency and blockchain communities, become a powerful governance mechanism. There are *project* forks, *code* forks and *chain* forks. In open-source development, forks are used so that people can work on parts of some code without disturbing the main code base. These forks can then be merged into the main code base as an addition, improvement, patch or otherwise – or they can go on to become a project of their own. Bitcoin has had thousands of code forks as people have sought to improve different aspects of the protocol and thousands of project forks as people have taken the Bitcoin protocol and modified it into new cryptocurrencies.[H, I, J] The early years of Bitcoin in particular saw a playful plethora of so-called 'altcoins', experiments trying out all kinds of different monetary, social and technical designs. A chain fork is related to these, but is different. Chain forks

---

H. See https://github.com/bitcoin/bips for how to contribute to Bitcoin development.
I. See https://github.com/bitcoin/bitcoin
J. See for example https://github.com/bitcoin/bitcoin/commits/master

don't only happen in reaction to contradictory blocks; they can also happen if miners decide to run a version of the Bitcoin client that might have within it a change that is incompatible with previous versions. It turns out that decentralisation can be coded in different ways. The Bitcoin scaling conflict, mentioned briefly earlier on, can be summarised broadly as being about how to technically accommodate for a growing number of transactions. The reference client has a hard data limit of 1MB per block. This was put in place early in the history of Bitcoin, to safeguard against micro payments being used as a denial-of-service (DoS) attack. Some wanted to increase the block size limit, while others argued that this would increase the data load and cause centralisation, arguing instead for developing faster layers on top of the existing protocol. It turned out that 'decentralisation' and 'scale' could be understood in different ways, representing different visions for the future of the Bitcoin project, benefiting some actors over others. Accusations of network attacks, CIA infiltration and compromised accounts as well as takeover attempts accumulated from 2016 onwards and resulted in a split. From this, several different Bitcoin protocols emerged, including Bitcoin Core, XT, UASF, Classic and Unlimited. Some of these are soft forks, meaning they are still compatible with earlier versions. Others, like Bitcoin Cash, are hard forks, meaning they are incompatible with earlier versions, forcing miners and nodes to decide which version of Bitcoin they believe is the best, most profitable or most compatible with their vision. Is this necessarily a bad thing? How should difference be accommodated for in open decentralised systems? Does decentralisation imply a single decentralised protocol? Or is decentralisation about multiple different protocols? Who benefits from forks and is forking friendly to people using the system?

New transaction broadcasts do not necessarily need to reach all nodes. As long as they reach many nodes, they will get into a block before long. Block broadcasts are also tolerant of dropped messages. If a node does not receive a block, it will request it when it receives the next block and realises it missed one.

6.   Incentive

By convention, the first transaction in a block is a special transaction that starts a new coin owned by the creator of the block.

Money creation (bitcoin mining 3/5) Automated, predictable and protocologically-determined money creation; this promise of Bitcoin has attracted people with many different critiques of existing money systems. Bitcoin mining serves as the money creation and distribution mechanism – in addition to being an anti-DoS mechanism and the method for verifying transactions and organising consensus. A new block will contain, amongst all the normal transactions, a kind of transaction that comes from the system itself and rewards the miner for their work. This is the money creation mechanism in Bitcoin. In the first years, a miner would receive 50 bitcoin per block success- fully mined. The reward is set to halve regularly as the network grows and more blocks are mined, to 25 bitcoin per block, then 12.5 bitcoin and so on, and will continue to be halved until rewards are stopped altogether. This is set to occur when a total of 21 million bitcoin are in circulation. At this point no new bitcoin will be created (apart from in the unlikely event of the network agreeing to run a new version of Bitcoin where this changes). The difficulty of mining and finding new blocks is regularly adjusted as more CPUs, GPUs and ASICs are added to keep block creation to an average of one block every ten minutes. Determining money creation through math- ematical difficulty in this way appeals to critics of existing money creation mechanisms. In right-wing economic thinking, the central bank controls money creation and can steal from populations by causing inflation. The idea is that limiting the money supply would ensure deflation instead of inflation. In the meantime, left-wing crit- icism of existing money creation tends to be aimed at private banks' accumulation of wealth by issuing debt. Because banks can lend far more money than they hold (also known as fractional reserve banking), the interest paid on those loans in effect creates more money, while central banks regulate such money creation by setting interest rates. For many, regardless of political background, the lure of the powerful protocol was a systems vision of money as a means to modify behaviour at scale. For anarchists of all colours, Bitcoin came to represent the possibility of programmable money, the power to experiment with and determine new economic and monetary rules that would remix market, social and political ideas in wild new designs.

> This adds an incentive for nodes to support the network, and provides a way to initially distribute coins into circulation, since there is no central authority to issue them. The steady addition of a constant amount of new coins is analogous to gold miners expending resources to add gold to circulation.

Gold (bitcoin mining 4/5) Some believe gold to be the truest form of money. The idea is that gold has an inherent value based on how rare it is and how difficult it is to acquire it. The gold analogy in Bitcoin suggests an affiliation with these ideas; that the value of a bitcoin token is derived from the effort that is needed to create it (or, more precisely, the burning of energy while repeatedly running hashing algorithms). The economic ideas contained in the gold analogy sound convincing from a peer-to-peer perspective: money that has an intrinsic value is in a sense *disintermediated* money, because it seems like it does not require an authority to back it up and enforce its value. This is also known as a commodity theory of money. The assumption is that an intrinsic natural value ensures stability, whereby the value is derived from the quality of the thing itself rather than being enforced and determined by a government and central bank. The alchemy of such ideas might be attractive, but the actual price of gold, as well as the value of bitcoins, are set in markets and determined purely by how much someone is willing to pay for them.

In our case, it is CPU time and electricity that is expended.

The incentive can also be funded with transaction fees. If the output value of a transaction is less than its input value, the difference is a transaction fee that is added to the incentive value of the block containing the transaction. Once a predetermined number of coins have entered circulation, the incentive can transition entirely to transaction fees and be completely inflation-free.

The incentive may help encourage nodes to stay honest.

Incentives (bitcoin mining 5/5) The work of mining is rewarded with the creation of new bitcoins. This little detail is explicitly designed to make it more profitable for someone to contribute to the network, by mining, than to attack it. With this simple idea, computer engineers and protocol developers were given an entirely new set of economic tools – and with this, the disruptive potential of Bitcoin became clear: the prospect of *a system that can pay for itself*. This represented, for many, the possibility to disrupt both the technical and economic models of existing internet monopolies and make them unfeasible. An internet infrastructure comprised and dominated by Google, Amazon and Facebook fundamentally relies on surveillance business models; large-scale data gathering for the purposes of fine-grained targeting (whether for advertisements or drone strikes, depending on who they sell the data to), a condition that some have given the name *surveillance capitalism*. Decentralised systems that prioritise privacy could disrupt such surveillance business models, but these have so far largely remained marginal. How would such infrastructures be funded at scale, and why would people contribute to the network? The economic incentives in Bitcoin allow for an internal economy to motivate and sustain open, decentralised contribution to the infrastructure. This is not just a nifty idea; it has the potential to fundamentally disrupt the way the entire internet operates economically, socially, politically and philosophically. However, the new decentralised business models of the future internet have yet to be developed, and as it turns out, inventing a token is not the same as establishing an economy.

If a greedy attacker is able to assemble more CPU power than all the honest nodes, he would have to choose between using it to defraud people by stealing back his payments or using it to generate new coins. He ought to find it more profitable to play by the rules, such rules that favour him with more new coins than everyone else combined, than to undermine the system and the validity of his own wealth.

Cryptoeconomics 'There is no way we are going to be able to model out every single thing', says Floersch, a developer from Ethereum, the second largest blockchain network after Bitcoin[K]. Floersch is working on a new consensus protocol, called 'proof-of-stake', with a modified cryptoeconomic design. The use of incentives in the Bitcoin protocol design gave rise to what is called cryptoeconomics. The field draws on ideas from game theory, economics, psychology and other disciplines that might offer tools for addressing behaviour in large systems through the design and implementation of systems with particular characteristics: specifically, to make bad behaviour expensive and good behaviour profitable and thereby desirable. The intention is to design economic incentives and punishments in such a way that the economic interest of the individual is aligned with that of the overall system. This proves much more complex than it sounds. On both a philosophical and practical level, the question of what constitutes 'the interest of the system' tends to be reduced to security concerns: how to prevent attacks and malicious behaviour. But what is malicious and what is honest behaviour in an open decentralised system? When is something an attack and when does it simply show a different understanding of how the system should or could work? The idea was that the system would be more neutral than potentially corrupt humans, but the very definition of good or bad behaviour brings with it assumptions about how things should work and for whom. Some protocol designers try and evade the responsibility of systems design decisions by claiming that

K.   Karl Floersch on Casper and proof-of-stake, https://youtu.be/ycF0WFHY5kc 11:55

cryptoeconomics is concerned with security alone, and in doing so remove themselves from politicised and socially consequential considerations. The task is to secure the network, and the economic rewards serve as security functions against DoS attacks. But there is no escaping the responsibility of the design decision. In cryptoeconomics, is the behaviour being engineered for the purposes of the system – or should the system be engineered for the purposes of specific uses?

Even considering security design issues alone, things quickly get complicated. Incorporating economic dynamics into the security model of a protocol opens up that protocol to all the potential economic dynamics that might affect it; this is particularly pronounced in systems that are being openly developed within public view. Because a token is not the same as an economy, the value of rewards received for mining depends on exchange rates; the profitability of mining depends on hardware and electricity costs as well as exchange rates; electricity costs depend on geographic location, policy and geopolitics; mining hardware depends on supply chains, labour costs; the exchange rates depend on demand but can also be manipulated by 'whales' who hold a lot of bitcoins, and so on. Drawing the lines around what matters in terms of security becomes a tricky task, and modeling everything that matters – well, as Floersch says, this is pretty much impossible. In such conditions, is it possible to model whether it is more profitable to contribute to the system than attack it? What do economic incentives really do in a decentralised system? At some point the concern of cryptoeconomics will have to shift from the needs of the system to that of the people, situations and relationships that a given system interfaces with, because here, with all the mushy humans, material realities and unknown beings is where meaning comes to matter in the first place.

7. Reclaiming Disk Space

Once the latest transaction in a coin is buried under enough blocks, the spent transactions before it can be discarded to save disk space. To facilitate this without breaking the block's hash, transactions are hashed in a Merkle Tree [7, 2, 5], with only the root included in the block's hash. Old blocks can then be compacted by stubbing off branches of the tree. The interior hashes do not need to be stored.

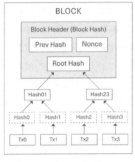

Transactions Hashed in a
Merkle Tree

After pruning Tx0-2 from
the Block

A block header with no transactions would be about 80 bytes. If we suppose blocks are generated every 10 minutes, 80 bytes * 6 * 24 * 365 = 4.2MB per year. With computer systems typically selling with 2GB of RAM as of 2008, and Moore's Law predicting current growth of 1.2GB per year, storage should not be a problem even if the block headers must be kept in memory.

8.   Simplified Payment Verification

It is possible to verify payments without running a full network node.

Full nodes Not all nodes are equal. The white paper already describes a 'simplified' node for those who do not want to hold the entire Bitcoin blockchain but still use the system to send transactions. In contrast, 'full' nodes are peers in the Bitcoin network that store the full Bitcoin blockchain and run a full Bitcoin client even though they do not necessarily mine. They participate by witnessing and relaying transactions, checking that the consensus rules (the Bitcoin reference client) are complied with. If a transaction is not in the correct format, or attempts something that is not permitted by the protocol, it is rejected by the full node peers and will not be added to the pool to be mined and verified. In this sense, full nodes are a first defence against attacks. Full nodes are different from miners, as they are not rewarded directly for their contribution in the network. This difference in motivation and reward is likely to shape the characteristics of these roles, and to a large extent has already done so; mining is increasingly done by mining pools (companies that have begun to resemble commercial service providers) whereas full nodes tend to be run by people looking to contribute on the basis of concern for the ethos of decentralisation and the overall development and governance of the project. The constitutional moment of Bitcoin, during the scaling conflict, brought about new understandings of the balances of power in open, decentralised systems. From this, full nodes decided to get organised to play an important role, using their capacity to witness and relay transactions to push through a protocol solution in the conflict. Through a campaign for a 'user activated soft fork' in 2017, people were encouraged to set up full nodes adopting a protocol change that would resolve the conflict through a compromise that would prevent the network from splitting (organised through #UASF). Full nodes can, in this way, participate in determining and implementing protocol changes by choosing which client, and thereby which consensus rules, to run. From nodes to simplified payment verification, lighting networks, mining and mining pools, wallets and more, the interactions with and ways of running and contributing to the network are emergent, far from uniform and still taking shape. Developers might implement protocol changes, but these need to be adopted by miners, agreed on by full nodes and used by people downloading wallets and so on. The governance of the Bitcoin protocol itself – indeed of open decentralised network protocols in general – is very much in development.

A user only needs to keep a copy of the block headers of the longest proof-of-work chain, which he can get by querying network nodes until he's convinced he has the longest chain, and obtain the Merkle branch linking the transaction to the block it's timestamped in. He can't check the transaction for himself, but by linking it to a place in the chain, he can see that a network node has accepted it, and blocks added after it further confirms the network has accepted it.

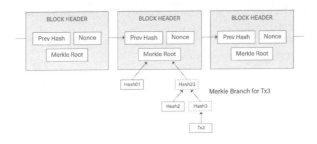

Merkle Branch for Tx3

As such, the verification is reliable as long as honest nodes control the network, but is more vulnerable if the network is overpowered by an attacker. While network nodes can verify transactions for themselves, the simplified method can be fooled by an attacker's fabricated transactions for as long as the attacker can continue to overpower the network. One strategy to protect against this would be to accept alerts from network nodes when they detect an invalid block, prompting the user's software to download the full block and alerted transactions to confirm the inconsistency. Businesses that receive frequent payments will probably still want to run their own nodes for more independent security and quicker verification.

9. Combining and Splitting Value

Although it would be possible to handle coins individually, it would be unwieldy to make a separate transaction for every cent in a transfer. To allow value to be split and combined, transactions contain multiple inputs and outputs. Normally there will be either a single input from a larger previous transaction or multiple inputs combining smaller amounts, and at most two outputs: one for the payment, and one returning the change, if any, back to the sender.

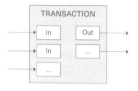

It should be noted that fan-out, where a transaction depends on several transactions and those transactions depend on many more, is not a problem here. There is never the need to extract a complete standalone copy of a transaction's history.

10. Privacy

The traditional banking model achieves a level of privacy by limiting access to information to the parties involved and the trusted third party. The necessity to announce all transactions publicly precludes this method, but privacy can still be maintained by breaking the flow of information in another place: by keeping public keys anonymous. **The public can see that someone is sending an amount to someone else, but without information linking the transaction to anyone. This is similar to the level of inform-ation released by stock exchanges, where the time and size of individual trades, the 'tape', is made public, but without telling who the parties were.**

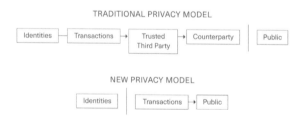

Transparency The Bitcoin blockchain is fully public, and the code of the Bitcoin protocol and clients are open-source. The system, in these senses, is transparent. Decentralised systems necessarily need to promise a certain level of transparency in order for them to be 'trustless' and secure. Yet there is a tension between transparency and privacy/anonymity. What should be transparent and what should be private? The Cypherpunks (a techno-political subculture that looks to leverage the power of cryptography to fight power) came up with a formula: transparency for the powerful and privacy for the powerless. When the powerful are the large internet corporations or the US government, this seems obvious. Leaking a document revealing what a government has done to civilians in a military operation uses transparency as a weapon against the powerful. It is specific, it is strategic, it is political; it is an act that entails a particular ethics. But what happens to transparency when it, just like decentralised networks, is no longer a specific weaponised strategy? Generalised transparency would be the same as generalised surveillance. Who, then, is the adversary exactly? Who is powerful and who is powerless within a decentralised system? What about wealthy individuals holding large amounts of bitcoins? What is considered important in terms of power: CPU power, amount of bitcoins, social graph or access to the reference client? And what about dollar wealth, social status, connections to policy makers and everything else that gives power to individuals? Who gets to decide? This is why transparency, privacy, decentralisation, openness and neutrality cannot be general values, but always entail specific decisions that are necessarily political. Some cryptographers are working on systems that selectively reveal only the necessary information in order to use a given system. Others are working on cryptographic proofs as a way to develop new accountability structures. There are, as of yet, no agreed upon methods to decide who is 'powerful' and who is 'powerless' in supposedly decentralised systems, who or what has the right to privacy and who or what should be transparent and open. These are not trivial questions.

As an additional firewall, a new key pair should be used for each transaction to keep them from being linked to a common owner. Some linking is still unavoidable with multi-input transactions, which necessarily reveal that their inputs were owned by the same owner. The risk is that if the owner of a key is revealed, linking could reveal other transactions that belonged to the same owner.

Privacy Systems, devices and technologies are an expression of what matters to those building and using them – and privacy matters to those who build cryptocurrency systems. To this point, so does anonymity. The anonymity and disappearance of Satoshi Nakamoto symbolised to many the disappearance of authority; the author of the system receded back into the darkness, the nym had served its purpose. Instead of coherent fixed identities, within Bitcoin the only fixed thing is the system, while all else is fluid, multiple and can exist in the dark, shielded from the prying eyes of authorities. This intention easily and quickly flips into its opposite. Research and testing has revealed several different ways that anonymity is compromised in Bitcoin: A given transaction can be traced all the way to an exchange and de-anonymised at this point. A bitcoin token is not an economy, so although accounts are not associated with names, most people look to cash out at some point, turning their bitcoins into dollars, yen, euros or their other preferred currency. In response, people are continuously developing new techniques and protocols to improve anonymity and privacy features. This is a fine art of selectively revealing and concealing, enabling dark spaces or selective floodlights – for example, coin mixers that can 'mix' transactions so that they cannot be traced directly to specific owners. Mix nets and advanced cryptography methods known as zero-knowledge proofs and the zk-SNARKS protocol have also been added to the arsenal for the development of different cryptocurrencies with much stronger privacy. Mechanisms such as Attribute Based Credentials (ABC) selectively reveal only the minimum necessary information in order be trusted and gain access to a system; they allow for a network of nyms that engage freely rather than necessitating coherent identities that can be targeted.[L] In response to those who seek complete determinacy, fixed identities and defined property, the fluid interactions of anonymous systems and privacy remains one of the more explicitly ethical and political-ly-aware areas of research and development in the fields of computer science and information security.

11. Calculations

We consider the scenario of an attacker trying to generate an alternate chain faster than the honest chain. Even if this is accomplished, it does not throw the system open to arbitrary changes, such as creating value out of thin air or taking money that never belonged to the attacker. Nodes are not going to accept an invalid transaction as payment, and honest nodes will never accept a block containing them. An attacker can only try to change one of his own transactions to take back money he recently spent.

The race between the honest chain and an attacker chain can be characterised as a Binomial Random Walk. The success event is the honest chain being extended by one block, increasing its lead by +1, and the failure event is the attacker's chain being extended by one block, reducing the gap by -1.

The probability of an attacker catching up from a given deficit is analogous to a Gambler's Ruin problem. Suppose a gambler with unlimited credit starts at a deficit and plays potentially an infinite number of trials to try to reach breakeven. We can calculate the probability he ever reaches breakeven, or that an attacker ever catches up with the honest chain, as follows[8]:

p = probability an honest node finds the next block
q = probability the attacker finds the next block
q_z = probability the attacker will ever catch up from z blocks behind

$$q_z = \left\{ \begin{array}{ll} 1 & if\, p \leq q \\ (q/p)^z & if\, p > q \end{array} \right\}$$

Given our assumption that $p > q$, the probability drops exponentially as the number of blocks the attacker has to catch up with increases. With

L. See Tim May's *The Crypto Anarchist Manifesto* and Eric Hughes' *A Cypherpunk Manifesto* for some concise explanations of privacy, power, networks and cryptography.

the odds against him, if he doesn't make a lucky lunge forward early on, his chances become vanishingly small as he falls further behind.

We now consider how long the recipient of a new transaction needs to wait before being sufficiently certain the sender can't change the transaction. We assume the sender is an attacker who wants to make the recipient believe he paid him for a while, then switch it to pay back to himself after some time has passed. The receiver will be alerted when that happens, but the sender hopes it will be too late.

The receiver generates a new key pair and gives the public key to the sender shortly before signing. This prevents the sender from preparing a chain of blocks ahead of time by working on it continuously until he is lucky enough to get far enough ahead, then executing the transaction at that moment. Once the transaction is sent, the dishonest sender starts working in secret on a parallel chain containing an alternate version of his transaction.

The recipient waits until the transaction has been added to a block and z blocks have been linked after it. He doesn't know the exact amount of progress the attacker has made, but assuming the honest blocks took the average expected time per block, the attacker's potential progress will be a Poisson distribution with expected value:

$$\lambda = z\,\frac{q}{p}$$

To get the probability the attacker could still catch up now [...]

Attacks Bitcoin is part of a culture and community that seeks to create a global network space free from what an eminent, politically-aware computer scientist has called formidable adversaries – a poetic way to describe the world's largest corporations along with the United States government and secret services more generally. The Bitcoin community is also very good at hacking the systems of the aforementioned formidable adversaries, and it is concerned with developing networks that might, in turn, withstand attacks by them. Understanding this, and looking at some of the main attacks on decentralised systems, gives some important insights into the concerns of those who build such systems. There are plenty of attack vectors, but two problems that any decentralised system faces are DDoS attacks (distributed denial-of-service) and Sybil attacks. A DDoS attack in Bitcoin could include, for example, flooding the network with lots of small transactions so that the network is so busy passing these around and verifying them that it effectively becomes clogged and can no longer be used for genuine transactions.[M] Sybil attacks occur when bots or other techniques are used to impersonate multiple nodes, while in reality, large parts of the network might be in the hands of only a few actors – essentially hidden authorities. There is an interesting and important dynamic here: because the intention is to create a network and an infrastructure that is beyond the influence of any authority, it is considered a neutral network, open for anyone to participate. What one developer of the second largest blockchain system called 'disinterested algorithmic interpreter' – that is, apolitical, and not influenced by the motivations of the people using the system. The system is considered a neutral substrate, and so those building it must only concern themselves with ensuring that the system functions as intended and is secure. For this reason, any problem or disagreement as to the functioning of the system tends to be addressed as a security question, rather than social, political, ethical or otherwise, and are solved by modelling and preventing attacks (whether by other hackers and by governments and corporations). There is a significant and deeply philosophical difference between claiming to build a neutral universal platform and building things with particular people and purposes in mind.

[...] ,we multiply the Poisson density for each amount of progress he could have made by the probability he could catch up from that point:

M. During the Bitcoin scaling conflict there were accusations that people arguing for larger block sizes were doing DDoS attacks on the network with transactions to make it seem like the network had reached capacity.

$$\sum_{k=0}^{\infty} \frac{\lambda^k e^{-\lambda}}{k!} \cdot \left\{ \begin{array}{ll} (q/p)^{(z-k)} & \textit{if } k \le z \\ 1 & \textit{if } k > z \end{array} \right\}$$

Rearranging to avoid summing the infinite tail of the distribution...

$$1 - \sum_{k=0}^{z} \frac{\lambda^k e^{-\lambda}}{k!} \left(1 - (q/p)^{(z-k)} \right)$$

Converting to C code...

```
#include
double AttackerSuccessProbability(double q, int z)
{
double p = 1.0 - q;
double lambda = z * (q / p);
double sum = 1.0;
int i, k;
for (k = 0; k <= z; k++)
{
    double poisson = exp(-lambda);
    for (i = 1; i <= k; i++)
        poisson *= lambda / i;
    sum -= poisson * (1 - pow(q / p, z - k));
}
return sum;
}
```

Running some results, we can see the probability drop off exponentially with z.

```
q=0.1
z=0    P=1.0000000
z=1    P=0.2045873
z=2    P=0.0509779
z=3    P=0.0131722
z=4    P=0.0034552
z=5    P=0.0009137
z=6    P=0.0002428
```

```
z=7      P=0.0000647
z=8      P=0.0000173
z=9      P=0.0000046
z=10     P=0.0000012


q=0.3
z=0      P=1.0000000
z=5      P=0.1773523
z=10     P=0.0416605
z=15     P=0.0101008
z=20     P=0.0024804
z=25     P=0.0006132
z=30     P=0.0001522
z=35     P=0.0000379
z=40     P=0.0000095
z=45     P=0.0000024
z=50     P=0.0000006
```

Solving for P less than 0.1%...

```
P < 0.001
q=0.10    z=5
q=0.15    z=8
q=0.20    z=11
q=0.25    z=15
q=0.30    z=24
q=0.35    z=41
q=0.40    z=89
q=0.45    z=340
```

12. Conclusion

We have proposed a system for electronic transactions without relying on trust. We started with the usual framework of coins made from digital signatures, which provides strong control of ownership, but is incomplete without a way to prevent double-spending. To solve this, we proposed a peer-to-peer network using proof-of-work to record a public history of transactions that quickly becomes computationally impractical for an attacker to change if

honest nodes control a majority of CPU power. The network is robust in
its unstructured simplicity. Nodes work all at once with little coordination.
They do not need to be identified, since messages are not routed to any
particular place and only need to be delivered on a best effort basis. Nodes
can leave and rejoin the network at will, accepting the proof-of-work chain
as proof of what happened while they were gone. They vote with their CPU
power, expressing their acceptance of valid blocks by working on extending
them and rejecting invalid blocks by refusing to work on them. **Any needed
rules and incentives can be enforced with this consensus mechanism.**

The truth machine And thus the truth machine was born.[N] This final
sentence of the white paper has been taken literally by a wave of new
blockchain projects and an emerging illuminati of believers; takfiris
debating and seeking to realise the true interpretation of the text.[O]
Since the invention of Bitcoin, its consensus mechanism (along with
the blockchain) has been generalised, moving from payment system
to platforms for decentralised computation and applied to any context
or condition, any set of rules. It produces a truth that is decided on
through probabilistically fair turns, determined by expending CPU
power and motivated by economic incentives. The truth that is arrived
at through the consensus mechanism lies not in deliberation, negotia-
tion, consensus of opinions or some notion of justice or objective truth,
but in randomness and in large numbers generating an operational,
computational consensus for the network. Where, previously, universal
truths have been imposed from above, from different gods, governments and ideologies, here it emerges from the consensus mechanism, from machines running algorithms, taking turns to determine
whether a transaction has happened or not and what a true event is.
This truth is a truth emerging through the probabilistic mediation of

N. Elsewhere, I have disassembled the particular form of 'truth' constructed here. See Brekke, J. K.
    Disassembling the Truth Machine. In: de Vega, M. Mazon Gardoqui, V. & Silvestrin, D. /META.
    Tracing unknown kno//wns/. ñ (Mexico City & Berlin).
O. Believers that interpret scriptures literally and take them as the only truth while everything else
    either needs to be converted or eliminated.

large numbers. To these believers, computational code is a disinterested and immediately executing law, decentralisation will solve the problem of power, cryptography will organise consensus and incentives will ensure good behaviour.

References

1. W. Dai, 'b-money,' http://www.weidai.com/bmoney.txt, 1998.
2. H. Massias, X.S. Avila, and J.-J. Quisquater, 'Design of a secure timestamping service with minimal trust requirements,' In *20th Symposium on Information Theory in the Benelux*, May 1999.
3. S. Haber, W.S. Stornetta, 'How to time-stamp a digital document,' In *Journal of Cryptology*, vol 3, no 2, pages 99-111, 1991.
4. D. Bayer, S. Haber, W.S. Stornetta, 'Improving the efficiency and reliability of digital time-stamping,' In *Sequences II: Methods in Communication, Security and Computer Science*, pages 329-334, 1993.
5. S. Haber, W.S. Stornetta, 'Secure names for bit-strings,' In *Proceedings of the 4th ACM Conference on Computer and Communications Security*, pages 28-35, April 1997.
6. A. Back, 'Hashcash - a denial of service counter-measure,'http://www.hashcash.org/papers/hashcash.pdf, 2002.
7. R.C. Merkle, "Protocols for public key cryptosystems," In Proc. 1980 *Symposium on Security and Privacy*, IEEE Computer Society, pages 122-133, April 1980.
8. W. Feller, "An introduction to probability theory and its applications," 1957.

The Bitcoin white paper was a proposal for a peer-to-peer payment system. In essence, for digital money. Yet the white paper lists solely cryptographic and computational references. There is not a single reference to any monetary, economic or political ideas and knowledge. Why? The proposition of Bitcoin is grounded in the field of decentralised computation. For those of you trying to understand and describe blockchain and cryptocurrencies as a strange subcultural anarcho-capitalism, speculation and financialisation on steroids or even a new frontier of authoritarian capitalist accumulation – stop. You have missed the point. It is not relevant; the territory has changed, the metrics look different and so do the political possibilities. These perspectives make for nice stories and there are elements of truth to them, but, as usual, there are far more interesting things happening than these traditional tales of never-ending profit extraction and large systems beyond our control. The machine has broken away. The boss no longer controls it.

Economic theories are now being subsumed and operationalised for the sake of computation, not the other way around. Classical economics and its cute little rational agents, who were once the only ones that 'counted', are becoming a marginal fiction, used for certain purposes in a much larger machine that looks towards higher goals. Economic, political, legal and social contexts and conditions are turned into computational and mathematical problems, solved in that realm, and then reapplied in an attempt at reconfiguring the economic, political, social and legal realms. This has caused both confusion and creativity; confusion by conflating computational models with political desires and social effects, and creativity by forcing a reassessment of the boundaries between disciplines, ideologies and entities. It has raised the question of which forces and sensibilities – mathematical, algorithmic, human, institutional, cultural or religious – we want to have determining which aspects of our lives.

What many blockchain believers and critics alike have missed is that this is no longer a world of capitalism looming and externalities struggling, no longer a battle for the marginal to make themselves matter and to be included in the equation. For too long the habit has been to think of politics as a singular expanding realm of 'things that matter', with externalities brought in to become counted. But now there are no more externalities. Now everything, potentially, matters.

Here, already, is a world of many, of differences, of multiple sensibilities, and what matters is how they relate. Because so many are still ignorant to this already existing multiplicity, the race in the bubbling industry of decentralisation is still, yet again, to become the protocol that will mediate difference and manage freedom with coordination, at global network scales. You know, The Final Platform; the one that does it all. Build any application, organisation, computation, any money-system, any capitalist/socialist/communist or anarchist system, any identity system, any legal system, anything. But build it on The Final Platform.

Bitcoin is this perfect '/trustless/ mathematical machine, built – most unfortunately – upon a foundation of mushy humans'; so wrote one person on the Bitcoin developers' mailing list.[P] Some look towards Bitcoin and decentralised computation as a single true reference point. If mathematics can be relied upon, then all else can be built on top. It can be an objective, deterministic system with which we can coordinate and organise all our differences and different perspectives, our subjective little selves roaming about.

And that, my dear friends, is a boring weak point. It is also the disappointment of the whole decentralisation movement. This is where Bitcoiners, blockchainers and their critics have all fallen embarrassingly short of understanding what is now at stake. Those of us who never lived with only one truth, one reality – we understand. Any project that strives for a unifying language, a substratum upon which the whole can be known, will always be a violent one, trying to impose its 'neutral' language, imprint its objective map and become 'the last cryptocurrency', as if there ever is a final arrival. Here come the Bitcoin maximalists, the blockchain believers, the computational takfiris. It doesn't matter if you call yourself a platform, a language, a protocol – no matter how deep you dig in the stack, the attempt to be the unifying neutral factor will always make you an enemy to those of us with two names. (The alliance, on the other hand, includes those eminently skilled engineers that understand that there is no innocence in systems design and therefore engage in creative collaboration across entities and sensibilities with generosity and humility.)

Bitcoin has succeeded in showing that money is not neutral; it can be programmed in many different ways. Now, after we have watched the protocol become generalised, it must be stated, and loudly so, that equally, computation is not neutral either. It is merely one sensibility, one method, out of many, for mediating difference.

P.   See https://lists.linuxfoundation.org/pipermail/bitcoin-dev/2014-April/005615.html

A few last words.

When I first read the Bitcoin white paper, I, like so many others, found it beautiful; a seemingly perfectly balanced combination of basic cryptography, simple economics and a decentralised network architecture. Decentralisation would take care of authority and power, economic incentives would solve security and behavioural problems and cryptography would secure the integrity of the system, its records and its use.

A curious double move happens when encountering such a convincing and elegant diagram: it begins to exist in our minds as an objective necessity, but at the same time it doesn't exist in reality quite so perfectly as it should, and we keep correcting, building, maintaining, reorganising the world around it to try and let it exist as it is, so elegant and real, in our minds. Here I don't mean that we correct for functionality, which is an honorable endeavor. What I mean, rather, is that we correct our understanding of the world so that it matches the model better. This is the seed of the takfiri. This is the definite determinacy so desperately sought after. The mind sees and understands the perfect diagram. It becomes a way to make sense of and order the world, a way to determine relationships, organise and enforce their terms and conditions once and for all. How easy it is, then, to dismiss everything else that is or might be.

And so he suddenly said, this young guy with dangly hair and nervous mouth at an Ethereum meetup in some city in Old Europa, 'anyone who cannot interact with blockchain and markets effectively perhaps is of inferior genetics and their extinction might not be a bad thing.' Alone, in a trustless world. With no communication with the world other than through systems diagrams, keys, wallets, markets and protocols and exchanges. Alone, with no way of knowing and relating to the world, save through ever more complex apparatuses. No other purpose than to optimise, hopefully and as quickly as possible beyond mushiness, towards that certain mathematical purity. But the system itself turned out to not be so pure, objective or easily separated from those who build, maintain and run it.

Determinism can be constructed internally for a particular arrangement in very creative and hugely useful ways. But the more that is brought into a deterministic relationship with such arrangements, the more this opens up a wider surface of indeterminacy. And a deterministic system is just another material phenomenon in the ongoing, emerging reality that we are part of. And then there are the rest of us, who have learned to be okay with the fact that the universe is so much stranger, more complex and indeterminate than we can ever imagine. We will never assume that we have understood and found the explanation for it all, or that we have built a perfect system to determine and fix everything – least of all through a systems diagram drawn up by one human (or a group of humans) in a specific political and economic context taking place over what does not even amount to a split second of history. It is, after all, just another way of being, thinking and doing in the world. It is not the first and most certainly won't be the last.

Bitcoin is an open decentralised system. Determinacy cannot be the end goal. Instead, here we are, living with indeterminacy, in the knowledge that the machine will never be finished. It will always benefit some over others, will always require corrections, maintenance or a complete remaking in order to accommodate for other sensibilities and new things that come to matter. Keep your head level, your heart open and choose your allies, your friends and your adversaries carefully; they come in surprising forms these days and do not necessarily take human, nor machinic, forms.

(And to systems takfiri of all kinds, a warning: it is a deadly mistake to see a system and its beautiful diagram as more real, objective and pure than everything else. Initially, perhaps, it is deadly for all those who will suffer from its (slow) violence. But later on, it is deadly for those who continue to perpetuate it, and who will therefore need to be stopped in order for the rest of us to breathe. Do not make the same mistake as market fundamentalists, colonialists, fascists and all the others who seek to fix the world in their image. Those of us with two names will eventually destroy you.)

Appendix

This selection of email correspondence, public postings and forum posts has been gathered together as reference material for the overall contents of this book. It should by no means be understood as exhaustive or fully representative of the vast and sprawling conversations that have taken place across a diverse array of platforms on the subject of Bitcoin.

Rather, this is a careful collation of historical textual material that supports and expands upon areas covered in this book. The material is presented here in chronological order from 2008 to 2011. For those wishing to delve deeper into the minutiae of material circulated in the public domain before and after the disappearance of Satoshi Nakamoto, the Satoshi Nakamoto Institute (https://nakamotoinstitute.org) has compiled a comprehensive database of all publically accessible information, as well as a library of historical texts from the forerunners of the crypto-anarchist and Cypherpunk movements as it relates to the creation of Bitcoin and more generally to cryptographically secure electronic cash.

Original Post On *The Cryptography Mailing List* 2008-11-01 19:16:33 UTC.

Bitcoin P2P e-cash paper

2008-11-01 19:16:33 UTC

I've been working on a new electronic cash system that's fully peer-to-peer, with no trusted third party.

The paper is available at:
http://www.bitcoin.org/bitcoin.pdf

The main properties:
Double-spending is prevented with a peer-to-peer network.
No mint or other trusted parties.
Participants can be anonymous.
New coins are made from Hashcash style proof-of-work.
The proof-of-work for new coin generation also powers the network to prevent double-spending.

Bitcoin: A Peer-to-Peer Electronic Cash System

Abstract. A purely peer-to-peer version of electronic cash would allow online payments to be sent directly from one party to another without the burdens of going through a financial institution. Digital signatures provide part of the solution, but the main benefits are lost if a trusted party is still required to prevent double-spending. We propose a solution to the double-spending problem using a peer-to-peer network. The network timestamps transactions by hashing them into an ongoing chain of hash-based proof-of-work, forming a record that cannot be changed without redoing the proof-of-work. The longest chain not only serves as proof of the sequence of events witnessed, but proof that it came

from the largest pool of CPU power. As long as honest
nodes control the most CPU power on the network,
they can generate the longest chain and outpace
any attackers. The network itself requires minimal
structure. Messages are broadcasted on a best effort
basis, and nodes can leave and rejoin the network at
will, accepting the longest proof-of-work chain as
proof of what happened while they were gone.

Full paper at:
http://www.bitcoin.org/bitcoin.pdf

Satoshi Nakamoto

The Cryptography Mailing List
Unsubscribe by sending '"unsubscribe cryptography"' to

Initial responses: is Bitcoin a political tool?

Source: The Cryptography Mailing List
http://satoshi.nakamotoinstitute.org/emails/cryptography/
4/#selection-7.0-11.23

Response 3

Unknown writes:

Re: Bitcoin P2P e-cash paper

Date: NA

> You will not find a solution to political problems
> in cryptography.

Satoshi writes:

2008-11-07 12:30:36 UTC

Yes, but we can win a major battle in the arms race and
gain a new territory of freedom for several years.

Governments are good at cutting off the heads of a
centrally controlled networks like Napster, but pure P2P
networks like Gnutella and Tor seem to be holding their
own.

Satoshi

The Cryptography Mailing List
Unsubscribe by sending '"unsubscribe cryptography'" to

Initial responses: Adding transaction to the blockchain.

Source: The Cryptography Mailing List
http://satoshi.nakamotoinstitute.org/emails/cryptography/5/

Response 5. Hal Finney writes:
Re: Bitcoin P2P e-cash paper
2008-11-09 14:13:34 UTC

> it is mentioned that if a broadcast transaction
> does not reach all nodes, it is OK, as it will get
> into the blockchain before long. How does this
> happen - what if the node that creates the '"next"'
> block (the first node to find the hashcash collision)
> did not hear about the transaction, and then a few
> more blocks get added also by nodes that did not
> hear about that transaction? Do all the nodes that
> did hear it keep that transaction around, hoping
> to incorporate it into a block once they get lucky
> enough to be the one which finds the next collision?

Satoshi writes:
Right, nodes keep transactions in their working set
until they get into a block. If a transaction reaches
90% of nodes, then each time a new block is found, it
has a 90% chance of being in it.

Hal Finney:
> Or for example, what if a node is keeping two or
> more chains around as it waits to see which grows
> fastest, and a block comes in for chain A which
> would include a double-spend of a coin that is in
> chain B? Is that checked for or not? (This might
> happen if someone double-spent and two different
> sets of nodes heard about the two different
> transactions with the same coin.)

Satoshi:

That does not need to be checked for. The transaction
in whichever branch ends up getting ahead becomes the
valid one, the other is invalid. If someone tries to
double- spend like that, one and only one spend will
always become valid, the others invalid.

Receivers of transactions will normally need to hold
transactions for perhaps an hour or more to allow time
for this kind of possibility to be resolved. They can
still re-spend the coins immediately, but they should
wait before taking an action such as shipping goods.

Hal Finney:
> I also don't understand exactly how double-spending,
> or cancelling transactions, is accomplished by
> a superior attacker who is able to muster more
> computing power than all the honest participants.
> I see that he can create new blocks and add them to
> create the longest chain, but how can he erase or
> add old transactions in the chain? As the attacker
> sends out his new blocks, aren't there consistency
> checks which honest nodes can perform, to make
> sure that nothing got erased? More explanation of
> this attack would be helpful, in order to judge the
> gains to an attacker from this, versus simply using
> his computing power to mint new coins honestly.

Satoshi:

The attacker isn't adding blocks to the end. He has to
go back and redo the block his transaction is in and
all the blocks after it, as well as any new blocks the
network keeps adding to the end while he's doing that.
He's rewriting history. Once his branch is longer, it
becomes the new valid one.

This touches on a key point. Even though everyone
present may see the shenanigans going on, there's no
way to take advantage of that fact.

It is strictly necessary that the longest chain is always considered the valid one. Nodes that were present may remember that one branch was there first and got replaced by another, but there would be no way for them to convince those who were not present of this. We can't have subfactions of nodes that cling to one branch that they think was first, others that saw another branch first, and others that joined later and never saw what happened. The CPU power proof-of-work vote must have the final say. The only way for everyone to stay on the same page is to believe that the longest chain is always the valid one, no matter what.

Hal Finney:
> As far as the spending transactions, what checks
> does the recipient of a coin have to perform?
> Does she need to go back through the coin's entire
> history of transfers, and make sure that every
> transaction on the list is indeed linked into the
> >'"timestamp'" block chain? Or can she just do the
> latest one?

Satoshi:
The recipient just needs to verify it back to a depth that is sufficiently far back in the block chain, which will often only require a depth of 2 transactions. All transactions before that can be discarded.

Hal Finney:
> Do the timestamp nodes check transactions, making
> sure that the previous transaction on a coin is
> in the chain, thereby enforcing the rule that all
> transactions in the chain represent valid coins?

Satoshi:
Right, exactly. When a node receives a block, it checks the signatures of every transaction in it against previous transactions in blocks. Blocks can only contain transactions that depend on valid transactions

in previous blocks or the same block. Transaction C could depend on transaction B in the same block and B depends on transaction A in an earlier block.

Hal Finney:

> Sorry about all the questions, but as I said this
> does seem to be a very promising and original idea,
> and I am looking forward to seeing how the concept
> is further developed. It would be helpful to see
> a more process oriented description of the idea,
> with concrete details of the data structures for
> the various objects (coins, blocks, transactions),
> the data which is included in messages, and
> algorithmic descriptions of the procedures for
> handling the various events which would occur in
> this system. You mentioned that you are working on
> an implementation, but I think a more formal, text
> description of the system would be a helpful next
> step.

Satoshi:

I appreciate your questions. I actually did this kind of backwards. I had to write all the code before I could convince myself that I could solve every problem, then I wrote the paper. I think I will be able to release the code sooner than I could write a detailed spec. You're already right about most of your assumptions where you filled in the blanks.

Satoshi Nakamoto

The Cryptography Mailing List
Unsubscribe by sending '"unsubscribe cryptography'" to

The Byzantine Generals' Problem.

Response 10: to James Donald, The Cryptography Mailing List 2008-11-13
22:34:25 UTC
Re: Bitcoin P2P e-cash paper 2008-11-13 22:34:25 UTC
James A. Donald wrote:
> It is not sufficient that everyone knows X. We
> also need everyone to know that everyone knows
> X, and that everyone knows that everyone knows
> that everyone knows X - which, as in the Byzantine
> Generals' problem, is the classic hard problem of
> distributed data processing.

The proof-of-work chain is a solution to the Byzantine
Generals' Problem. I'll try to rephrase it in that
context.

A number of Byzantine Generals each have a computer
and want to attack the King's wi-fi by brute forcing the
password, which they've learned is a certain number of
characters in length. Once they stimulate the network
to generate a packet, they must crack the password
within a limited time to break in and erase the logs,
otherwise they will be discovered and get in trouble.
They only have enough CPU power to crack it fast enough
if a majority of them attack at the same time.

They don't particularly care when the attack will
be, just that they all agree. It has been decided
that anyone who feels like it will announce a time,
and whatever time is heard first will be the official
attack time. The problem is that the network is not
instantaneous, and if two generals announce different
attack times at close to the same time, some may hear
one first and others hear the other first.

They use a proof-of-work chain to solve the problem. Once each general receives whatever attack time he hears first, he sets his computer to solve an extremely difficult proof-of-work problem that includes the attack time in its hash. The proof-of-work is so difficult, it's expected to take 10 minutes of them all working at once before one of them finds a solution. Once one of the generals finds a proof-of-work, he broadcasts it to the network, and everyone changes their current proof-of-work computation to include that proof-of-work in the hash they're working on. If anyone was working on a different attack time, they switch to this one, because its proof-of-work chain is now longer.

After two hours, one attack time should be hashed by a chain of 12 proofs-of-work. Every general, just by verifying the difficulty of the proof-of-work chain, can estimate how much parallel CPU power per hour was expended on it and see that it must have required the majority of the computers to produce that much proof-of-work in the allotted time. They had to all have seen it because the proof-of-work is proof that they worked on it. If the CPU power exhibited by the proof-of-work chain is sufficient to crack the password, they can safely attack at the agreed time.

The proof-of-work chain is how all the synchronisation, distributed database and global view problems you've asked about are solved.

The Cryptography Mailing List
Unsubscribe by sending '"unsubscribe cryptography'" to

On libertarianism.

Response 11: to Hal Finney, The Cryptography Mailing List 2008-11-14 17:29:22 UTC

Re: Bitcoin P2P e-cash paper 2008-11-14 17:29:22 UTC

Hal Finney wrote:

> I think it is necessary that nodes keep a separate
> pending-transaction list associated with each
> candidate chain. ... One might also ask ... how many
> candidate chains must a given node keep track of at
> one time, on average?

Fortunately, it's only necessary to keep a pending-transaction pool for the current best branch. When a new block arrives for the best branch, ConnectBlock removes the block's transactions from the pending-tx pool. If a different branch becomes longer, it calls DisconnectBlock on the main branch down to the fork, returning the block transactions to the pending-tx pool, and calls ConnectBlock on the new branch, sopping back up any transactions that were in both branches. It's expected that reorgs like this would be rare and shallow.

With this optimisation, candidate branches are not really any burden. They just sit on the disk and don't require attention unless they ever become the main chain.

> Or as James raised earlier, if the network
> broadcast is reliable but depends on a potentially
> slow flooding algorithm, how does that impact
> performance?

Broadcasts will probably be almost completely reliable. TCP transmissions are rarely ever dropped these days, and the broadcast protocol has a retry mechanism to get the data from other nodes after a while. If broadcasts turn out to be slower in practice than expected, the target time between blocks may have to be increased to avoid wasting resources. We want blocks to usually propagate in much less time than it takes to generate them, otherwise nodes would spend too much time working on obsolete blocks.

I'm planning to run an automated test with computers randomly sending payments to each other and randomly dropping packets.

> 3. The bitcoin system turns out to be socially
> useful and valuable, so that node operators feel
> that they are making a beneficial contribution to
> the world by their efforts (similar to the various
> '"@Home'" compute projects where people volunteer
> their compute resources for good causes).
>
> In this case it seems to me that simple altruism
> can suffice to keep the network running properly.

It's very attractive to the libertarian viewpoint if we can explain it properly. I'm better with code than with words though.

Satoshi Nakamoto

The first release of Bitcoin on sourceforge.net.

Response 4: The Cryptography Mailing List 2009-01-09 20:05:49 UTC
Bitcoin v0.1 released 2009-01-09 20:05:49 UTC
Announcing the first release of Bitcoin, a new electronic cash system that uses a peer-to-peer network to prevent double-spending. It's completely decentralisized with no server or central authority.

See bitcoin.org for screenshots.

Download link:
http://downloads.sourceforge.net/bitcoin/bitcoin-0.1.0.rar

Windows only for now. Open source C++ code is included.

- Unpack the files into a directory
- Run BITCOIN.EXE
- It automatically connects to other nodes

If you can keep a node running that accepts incoming connections, you'll really be helping the network a lot. Port 8333 on your firewall needs to be open to receive incoming connections.

The software is still alpha and experimental. There's no guarantee the system's state won't have to be restarted at some point if it becomes necessary, although I've done everything I can to build in extensibility and versioning.

You can get coins by getting someone to send you some, or turn on Options->Generate Coins to run a node and generate blocks. I made the proof-of-work difficulty ridiculously easy to start with, so for a little while in the beginning a typical PC will be able to generate coins in just a few hours. It'll get a lot harder when

competition makes the automatic adjustment drive up
the difficulty. Generated coins must wait 120 blocks to
mature before they can be spent.

There are two ways to send money. If the recipient is
online, you can enter their IP address and it will
connect, get a new public key and send the transaction
with comments. If the recipient is not online, it is
possible to send to their Bitcoin address, which is a
hash of their public key that they give you. They'll
receive the transaction the next time they connect and
get the block it's in. This method has the disadvantage
that no comment information is sent, and a bit of
privacy may be lost if the address is used multiple
times, but it is a useful alternative if both users
can't be online at the same time or the recipient
can't receive incoming connections.

Total circulation will be 21,000,000 coins. It'll be
distributed to network nodes when they make blocks,
with the amount cut in half every 4 years.

first 4 years: 10,500,000 coins
next 4 years: 5,250,000 coins
next 4 years: 2,625,000 coins
next 4 years: 1,312,500 coins
etc...

When that runs out, the system can support transaction
fees if needed. It's based on open market competition,
and there will probably always be nodes willing to
process transactions for free.

Satoshi Nakamoto

The Cryptography Mailing List
Unsubscribe by sending '"unsubscribe cryptography'" to

Satoshi rewrites the original Bitcoin statement for P2P Foundation website.

Bitcoin open- source implementation of P2P currency
P2P Foundation
2009-02-11 22:27:00 UTC
I've developed a new open- source P2P e-cash system
called Bitcoin. It's completely decentraliszed, with no
central server or trusted parties, because everything
is based on crypto proof instead of trust. Give it a
try, or take a look at the screenshots and design paper:

Download Bitcoin v0.1 at http://www.bitcoin.org

The root problem with conventional currency is all
the trust that's required to make it work. The central
bank must be trusted not to debase the currency, but
the history of fiat currencies is full of breaches of
that trust. Banks must be trusted to hold our money
and transfer it electronically, but they lend it out
in waves of credit bubbles with barely a fraction in
reserve. We have to trust them with our privacy, trust
them not to let identity thieves drain our accounts.
Their massive overhead costs make micropayments
impossible.

A generation ago, multi-user time-sharing computer
systems had a similar problem. Before strong encryption,
users had to rely on password protection to secure
their files, placing trust in the system administrator
to keep their information private. Privacy could always
be overridden by the admin based on his judgment
call, weighing the principle of privacy against other
concerns, or at the behest of his superiors. Then
strong encryption became available to the masses, and
trust was no longer required. Data could be secured

in a way that was physically impossible for others to access, no matter for what reason, no matter how good the excuse, no matter what.

It's time we had the same thing for money. With e-currency based on cryptographic proof, without the need to trust a third party middleman, money can be secure and transactions effortless.

One of the fundamental building blocks for such a system is digital signatures. A digital coin contains the public key of its owner. To transfer it, the owner signs the coin together with the public key of the next owner. Anyone can check the signatures to verify the chain of ownership. It works well to secure ownership, but leaves one big problem unsolved: double-spending. Any owner could try to re-spend an already spent coin by signing it again to another owner. The usual solution is for a trusted company with a central database to check for double-spending, but that just gets back to the trust model. In its central position, the company can override the users, and the fees needed to support the company make micropayments impractical.

Bitcoin's solution is to use a peer-to-peer network to check for double-spending. In a nutshell, the network works like a distributed timestamp server, stamping the first transaction to spend a coin. It takes advantage of the nature of information being easy to spread but hard to stifle. For details on how it works, see the design paper at http://www.bitcoin.org/bitcoin.pdf

The result is a distributed system with no single point of failure. Users hold the crypto keys to their own money and transact directly with each other, with the help of the P2P network to check for double-spending.

Satoshi Nakamoto
http://www.bitcoin.org

Email exchanges between Satoshi Nakamoto and Mike Hearn.

During this period, the soon-to-be core developer Mike Hearn questions Satoshi Nakamoto over email regarding a number of specific decisions that were taken in the design of the Bitcoin protocol.

Source: https://pastebin.com/Na5FwkQ4

Satoshi writes (In response to Mike):
My choice for the number of coins and distribution schedule was an educated guess. It was a difficult choice, because once the network is going it's locked in and we're stuck with it. I wanted to pick something that would make prices similar to existing currencies, but without knowing the future, that's very hard. I ended up picking something in the middle. If Bitcoin remains a small niche, it'll be worth less per unit than existing currencies. If you imagine it being used for some fraction of world commerce, then there's only going to be 21 million coins for the whole world, so it would be worth much more per unit. Values are 64-bit integers with 8 decimal places, so 1 coin is represented internally as 100000000. There's plenty of granularity if typical prices become small. For example, if 0.001 is worth 1 Euro, then it might be easier to change where the decimal point is displayed, so if you had 1 Bitcoin it's now displayed as 1000, and 0.001 is displayed as 1.

More email correspondence here:
https://pastebin.com/Na5FwkQ4
https://pastebin.com/cKZPC1rF
https://pastebin.com/wA9Jn100
https://pastebin.com/JF3USKFT
https://pastebin.com/syrmi3ET

A discussion of a proposed deletion of the Bitcoin Wikipedia entry.

Bitcointalk
2010-07-20 18:38:28 UTC
Giulio Prisco, July 14, 2010:
http://en.wikipedia.org/wiki/Bitcoin

This article is being considered for deletion in accordance with Wikipedia's deletion policy.
Please share your thoughts on the matter at this article's entry on the Articles for deletion page.

This article needs references that appear in reliable third-party publications. Primary sources or sources affiliated with the subject are generally not sufficient for a Wikipedia article. Please add more appropriate citations from reliable sources.

The recent Slashdot article should be considered as a reliable reference: http://news.slashdot.org/story/10/07/11/1747245/Bitcoin-Releases-Version-03

I cannot edit at this moment, can you guys save the WP articleartcile?

Note: Mt. Gox launches on July 18, 2010 https://en.wikipedia.org/wiki/Mt._Gox (Digital Gold p. 51)

Satoshi, July 20, 2010:
Bitcoin is an implementation of Wei Dai's b-money proposal http://weidai.com/bmoney.txt on Cypherpunks http://en.wikipedia.org/wiki/Cypherpunks in 1998 and Nick Szabo's Bitgold proposal http://unenumerated.blogspot.com/2005/12/bit-gold.html

The timing is strange, just as we are getting a rapid increase in third3rd party coverage after getting slashdotted. I hope there's not a big hurry to wrap the discussion and decide. How long does Wikipedia typically leave a question like that open for comment?

It would help to condense the article and make it less promotional sounding as soon as possible. Just letting people know what it is, where it fits into the electronic money space, not trying to convince them that it's good. They probably want something that just generally identifies what it is, not tries to explain all about how it works.

If you post in http://en.wikipedia.org/wiki/Wikipedia:Articles_for_deletion/ Bitcoin please don't say '"yeah, but bitcoin is really important and special so the rules shouldn't apply'" or argue that the rule is dumb or unfair. That only makes it worse. Try to address how the rule is satisfied.

Search '"bitcoin'" on Ggoogle and see if you can find more big references in addition to the infoworld and slashdot ones. There may be very recent stuff being written by reporters who heard about it from the slashdot article.

I hope it doesn't get deleted. If it does, it'll be hard to overcome the presumption. Institutional momentum is to stick with the last decision. (edit: or at least I assume so, that's how the world usually works, but maybe Wiki is different)

On July 31st, the article was officially deleted and then later restored.

Public post on *Bitcointalk* regarding the energy efficiency of Bitcoin.

Bitcoin minting is thermodynamically perverse
August 06, 2010, 01:52:00 AM

gridecon
Jr. Member

Let me begin by saying that Bitcoin is an amazing project and I am very impressed with the implementation and the goals. From reading these forums it seems to be understood that debate about the design and operation of the bitcoin economy ultimately serves to strengthen it, so I hope these comments are taken in that spirit. *EDIT - I have been convinced by further research and discussion that Bitcoin is actually highly efficient compared to most traditional currencies, because the infrastructure required to support a government issued fiat currency represents a much larger investment of resources than Bitcoin's CPUcpu power consumption. I am leaving this thread active though because it has been generating a lot of interesting discussion.*

I believe that the amount of energy input required to the bitcoin economy represents a serious obstacle to its growth. I think in the long-term, transactions may be even more serious than minting in this regard, but I will for the moment discuss minting because it is more precisely bounded and defined. The idea that the value of bitcoins is in some way related to the value of the electricity required, on average, to mint a winning block is generally accepted, but the precise nature of this relationship is contentious.

One argument is that anyone who chooses to generate coins is actually making the choice to purchase bitcoins with electricity/computational resources, and that because some/many people are in fact making that choice, bitcoins have at least that much '"value"' to the generators, who can be assumed to be maximizing their utility. A contrasting argument is that cost of production is different than market value, and the most objective measure is the current market conversion price to a more liquid and widely traded currency such as the US dollar.

My contention is that both of these arguments miss the point and the real problem, which is the fundamental perversity of wasting large amounts of energy and computations in generating the winning blocks for the minting process. The minting process exists because of the necessity of actually '"printing"' the currency, and certain desirable properties of crypto-math for making the currency's behavior predictable. The fact that the current minting process requires a large energy input of computational work is highly unfortunate and has the perverse consequence that bitcoin may actually be '"destroying wealth"' in the sense of wasting energy producing a digital object worth less than the resources invested in it.

As is often pointed out, a currency does not necessarily have, or need to have, any inherent value - a medium of exchange is a useful tool and can have value purely as a consequence of social convention. The cost of production of bitcoins in electricity consumed represents a waste, a '"thermodynamic burden"' that the currency has to carry. Consider a hypothetical alternative digital currency called '"compucoin"', which purchases cpu cycles from nodes on the network. The market value of this currency would converge very closely with the cost of electricity required to generate cpu cycles. Instead of costing cpu cycles to mint, the value of the cpu cycles the coins could be exchanged for would create a rational basis for the currency's value and integrate it with an existing market. I imagine that alternatives to Bitcoin (many of them probably sharing a lot of Bitcoin's source code) will inevitably emerge and Bitcoin's current minting process makes the currency '"expensive"' in terms of energy input. I believe this places it at a competitive disadvantage to other currencies and can only hinder its widespread adoption and long-term value. *Edit - as mentioned above, I am now much more optimistic about Bitcoin long term. I still think compucoins would be a cool idea, though!*

See here for discourse between forum members:
https://bitcointalk.org/index.php?topic=721.msg7889#msg7889

Satoshi's response:
August 07, 2010, 05:46:09 PM
It's the same situation as gold and gold mining. The marginal cost of gold mining tends to stay near the price of gold. Gold mining is a waste, but that waste is far less than the utility of having gold available as a medium of exchange.

I think the case will be the same for Bitcoin. The utility of the exchanges made possible by Bitcoin will far exceed the cost of electricity used. Therefore, not having Bitcoin would be the net waste.

Each node's influence on the network is proportional to its CPU power. The only way to show the network how much CPU power you have is to actually use it.

If there's something else each person has a finite amount of that we could count for one-person-one-vote, I can't think of it. IP addresses... much easier to get lots of them than CPUs.

I suppose it might be possible to measure CPU power at certain times. For instance, if the CPU power challenge was only run for an average of 1 minute every 10 minutes. You could still prove your total power at given times without running it all the time. I'm not sure how that could be implemented though. There's no way for a node that wasn't present at the time to know that a past chain was actually generated in a duty cycle with 9 minute breaks, not back to back.

Proof-of-work has the nice property that it can be relayed through untrusted middlemen. We don't have to worry about a chain of custody of communication. It doesn't matter who tells you a longest chain, the proof-of-work speaks for itself.

Satoshi later:
2010-08-09 21:28:39 UTC

The heat from your computer is not wasted if you need to heat your home. If you're using electric heat where you live, then your computer's heat isn't a waste. It's equal cost if you generate the heat with your computer.

If you have other cheaper heating than electric, then the waste is only the difference in cost.

If it's summer and you're using A/C, then it's twice.

Bitcoin generation should end up where it's cheapest. Maybe that will be in cold climates where there's electric heat, where it would be essentially free.

The Wikileaks Issue.

The community on Bitcointalk discusses reaching out to Wikileaks to offer Bitcoin as a solution to funds being seized in 2010 just before Paypal withdraws services (Dec 2004). The conversation that follows leads to discussion on Julian Assange, democracy, the legalities of Bitcoin and a debate concerning a strategy for the mainstream adoption of Bitcoin.

Once Satoshi voices dissent, the forum follows. Finally, PC World publishes a speculative article about Wikileaks and Bitcoin which ties the two together in the public's conscience. Many have speculated that this is what caused Satoshi to step away from the Bitcoin project.

Source: Bitcointalk.org
Full conversation: https://bitcointalk.org/index.php?topic=1735.
msg26999#msg26999
Excerpt of conversation below:

Genjix writes:
Wikileaks contact info?
November 10, 2010, 12:49:16 PM

> I wanted to send a letter to Wikileaks about Bitcoin since unfortunately they've had several incidents where their funds have been seized in the past.
> http://wikileaks.org/media/support.html
> Anyone know where to send a message to them?

ShadowOfHarbringer writes:
Re: Wikileaks contact info?
November 10, 2010, 01:28:00 PM

> Yeah this may be good for wikileaks, but not necessarily good for Bitcoin. If bitcoin becomes publicly associated with wikileaks before going semi-mainstream, then it will be viewed by the 'ruling class' (read: US government) as a tool for doing money laundering and another 'gray' businesses.

This may make bitcoin be considered by governments as a serious threat, and they will start fighting it too soon, holding back mainstream adoption.

Of course i know that they (governments) will start fighting bitcoin at some point, but the later, the better. If we go mainstream first, then such efforts will be probably futile, and BTC will probably be unstoppable without shutting down whole internet.

Anonymous writes:
Re: Wikileaks contact info?
November 10, 2010, 02:18:22 PM

Contrary to popular opinion it is not illegal to create your own currency. If they try to claim there is something illegal in the mere act of transferring value or bits of data they will look ridiculous.

Sturle writes:
Re: Wikileaks contact info?
November 10, 2010, 02:12:42 PM

By your logic the USA should forbid money. At least on paper and metal, a lot of currencies, and money transfers to uncontrollable countries like Bermuda and Switzerland. Perhaps go completely communist. On a worldwide basis porn and gambling is a lot more controversial than Wikileaks. Only China and Thailand censor Wikileaks. A lot more countries censor porn and gambling sites. I don't think this will make the US government blink...

MoonShadow writes:
Re: Wikileaks contact info?
November 11, 2010, 12:03:40 AM

Quote from: BitLex on November 10, 2010, 11:15:51 PM
Quote from: ShadowOfHarbringer on November 10, 2010, 01:28:00 PM
This may make bitcoin be considered by governments as a serious threat, and they will start fighting it too soon, ...

if they gonna fight it anyway, isn't it better the sooner they do? or do you want to invest (time/money/effort) into it first and have them fight it later? The later the better. The longer that governments wait to act, the stronger the bitcoin network grows, and the harder it becomes to harm it.

ShadowOfHarbringer writes:
Re: Wikileaks contact info?
November 11, 2010, 04:24:58 PM

They can't shoot bitcoin or bittorrent itself, but they can discourage big businessesbuisnesses from accepting bitcoin. And that can easily hold mainstream adoption even for tens of years.

They (governments) are very good in this particularly. Large companies such as Google, Microsoft, Wal-mart etc are very susceptible to government regulations.

Mikegogulski writes:
Re: Wikileaks contact info?
November 29, 2010, 01:01:53 PM

Screw big business. Google, Microsoft and Wal-Mart can all eat flaming death as far as I'm concerned, creatures of state privilege as they are.

My task, and where systems like Bitcoin can be helpful, is in making both government and big business irrelevant and obsolete.

Burn, Hollywood, burn.

Conversation continues on the subject of Julian Assange.

(see: https://bitcointalk.org/index.php?topic=1735.80)

mikegogulski writes:
Re: Wikileaks contact info?
November 29, 2010, 02:24:48 PM

> I believe very strongly that a successful effort to supplant state money is going to be a ground-up effort. The big players in today's pink economy are going to be the last to adapt, kicking and screaming, to the new economic reality. For years before that happens, though, individuals, sole traders and small enterprises are going to be taking more and more of their income away as, simultaneously, a) the new monetary system's benefits impact more and more people, b) the inevitable institutional dumbness of big organizations makes them stumble and lag, and, c) continually eroding confidence in the state, its institutions and the entities that rely upon them drives more and more people into the new economy.
> If you're reading this, you're part of the vanguard building the new economy and the new world. Mainstream acceptance will come by weight of numbers and via network effects. Who ever heard of Wikileaks or Twitter four years ago? And today we've got Admiral Chairman-of-the-Joint-Fucking-Chiefs Mullen tweeting his dismay over Wikileaks -- no doubt via the intermediation of an office full of PR analysts and bureaucrats, at extremely low relative impact.
> BTW, Shadow, when they shut down the internet, I know how to build a new one. Want to help?

Wumpus writes:
Re: Wikileaks contact info?
December 04, 2010, 08:47:59 AM

> Paypal just blocked them, and they're trying to get other US banks do the same. This would be a great moment to open bitcoin donations.

S3052 writes:
Re: Wikileaks contact info?
December 04, 2010, 08:57:41 AM

as much as what's going on with wikileaks is not right,
do we really think it makes sense to push bitcoins into wikileaks is making sense in the current environment?
i am scared that this poses a big risk for bitcoins to consciously go against too many and too big enemies

It would potentially be much better to first make bitcoin be far more widespread and accepted in a safe way vs going against the big ones already now

M0mchil
Re: Wikileaks contact info?
December 04, 2010, 09:27:21 AM
Great danger? But wait, isn't bitcoin invincible?! (Well, perhaps if it adopts random ports, protocol obfuscation, DHT bootstrapping...)

Jgarzik writes:
Re: Wikileaks contact info?
December 04, 2010, 09:59:27 AM
+1, I agree completely

We know that private and government forces are actively tracing, and trying to shut down, sources of wikileaks funding through all available means of pressure.

Does it make sense to actively give multiple world governments incentive to shut down bitcoin?

No matter how sympathetic wikileaks' cause... if you care about bitcoin's success, the answer is no.

Bruce Wagner writes:
This is a very good point.

'"PayPal Freezes WikiLeaks Account"'   http://www.wired.com/threatlevel/2010/12/paypal-wikileaks/

We must weigh...

---> the potential financial benefit bitcoin would give to wikileaks at this moment

vs.

---> the potential demise of bitcoin itself (or at least the impact of bitcoin being mortally wounded at this very early stage of its existence.)

I think that -- no matter how righteous anyone may feel about the work wikileaks does...

1. They don't need that much money to do what they do.
2. What they could potentially get via bitcoin at the moment wouldn't be a drop in the bucket of what they're receiving via other private means.
3. Nothing wikileaks is doing is WORTH the potential demise of Bitcoin.

We don't need to make Bitcoin, or each of US for that matter, into targets of world governments.   Especially not so early in Bitcoin's life.

Wikileaks is probably being funded, at least partially, by other governments with interests in leaking damaging information.

But in any case, Wikileaks probably has a bankroll larger than the ENTIRE bitcoin economy at this point in time.

Also, when you create a new Iron Man suit in your lab, you don't take on all the militaries of the world on your first trip out on the '"test track'". Bitcoin is still in '"beta'", remember.

I say, we MUST get Bitcoin accepted at Starbucks and the local grocery store.... BEFORE it gets accepted at Wikileaks.

Then, we'll have a chance.

Starbucks.   Anyone have contact info for them?

Hal Writes:
Re: Wikileaks contact info?
December 04, 2010, 08:43:07 PM

Looking on the bright side, if Bitcoin did get known as the Wikileaks currency, attacked by governments all over the world, at least we'd get our Wikipedia page back! [In reference to Wikipedia taking down their page]

Sturle writes:
Re: Wikileaks contact info?
December 04, 2010, 09:38:29 PM

Wikileaks have not stolen any secrets or signed any confidentiality agreements. They are just printing documents given to them by other people. The people who gave the secret documents to Wikileaks were probably doing something illegal. Not Wikileaks, or the newspapers which printed the leaked documents before they were available from Wikileaks. So, why aren't the editors of the newspapers arrested, threatened to be murdered, getting their servers and DNS shut down., accounts closed, etc? Because what they are doing is perfectly legal, and so is Wikileaks. Just unpopular among some who believes the press should write what the government tells them to write.

RHorning writes:
Re: Wikileaks contact info?
December 04, 2010, 10:17:44 PM

This is so true. There certainly wouldn't be a shortage of '"reliable sources'" about Bitcoins at that point. I think it would likely show up on the front page of most newspapers and be talked about extensively on both radio talk shows and the other broadcast networks too.

For myself, I'm getting to the point to say '"bring it on'" in regards to Wikileaks. Note that I'm using my real name here instead of a pseudonymp-suedonym and I'm willing to personally say '"bring it on'" in terms of being associated with Bitcoins as a project. I've had police come into my house without my permission already and do all kind of stupid stuff, so for me that line being crossed has already happened. I am also connected to enough people politically that if something was to happen to me that it would be noted and things would happen too.

It is the morally correct thing to be supporting Wikileaks, and if they'll take a few of my bitcoins, I not only want to donate but to let the world know that they can donate to Wikileaks through Bitcoins as well.

I can't speak for everybody here in the Bitcoins community but I am speaking for myself on this matter, and I'm not afraid of anything that the U.S. government might do to me if I was associated with backing Wikileaks financially. If anything, it would show that I no longer live under a constitutional government any more. If the U.S. government wants to tip their hand to expose themselves in that way, so be it. If the U.S. government kills me or puts me in jail, I'll certainly set a way for this community to find out. I really don't think it would come to that either, but I don't care if it did.

If I have to '"vote"' on this matter, I would encourage the Bitcoin community to take up the plate like we did with the EFF and encourage Wikileaks to put up a Bitcoin address on their website for donations. It would bring in some new blood into the Bitcoin community regardless, and it might be beneficial to Wikileaks as well. Leave it to Wikileaks to see if they want to use Bitcoins or not. In terms of governmental review of Bitcoins, we know that is going to happen sooner or later, so why are we fighting that inevitable result? Anything other than a low-key investigation is only going to make more people interested in Bitcoins, which is only going to help the project even more. It can't be killed as a project, only slowed down a little bit in its growth at this point and more likely its adoption would be accelerated by any kind of publicity that would happen.

The only possible concern I would have is over how sound the protocol itself is right now. If anything, a major flux of new people into Bitcoins would help there too, and the worst that could happen is that Bitcoins itself would be broken in some way where a new cryptocurrency would have to be created fixing the problems of Bitcoins. It is the idea of cryptocurrency that would then persist, and it is incredibly hard to censor an idea.

Basically, bring it on. Let's encourage Wikileaks to use Bitcoins and I'm willing to face any risk or fallout from that act.

-- Robert S. Horning
Logan, Utah

Conversation continues back and forth.

Satoshi writes:
Re: Wikileaks contact info?
December 05, 2010, 09:08:08 AM

> *Quote from: RHorning on December 04, 2010, 10:17:44 PM*
> Basically, bring it on. Let's encourage Wikileaks to use Bitcoins and I'm
> willing to face any risk or fallout from that act.

No, don't '"bring it on'".

The project needs to grow gradually so the software can be strengthened
along the way.

I make this appeal to WikiLeaks not to try to use Bitcoin. Bitcoin is a small
beta community in its infancy. You would not stand to get more than
pocket change, and the heat you would bring would likely destroy us at
this stage.

Following Satoshi's post the view of the community shifts.

Da2ce7 writes:
Re: Wikileaks contact info?
December 05, 2010, 02:17:40 PM

> *Quote from: satoshi on December 05, 2010, 09:08:08 AM*
>> *Quote from: RHorning on December 04, 2010, 10:17:44 PM*
>> *Basically, bring it on. Let's encourage Wikileaks to use Bitcoins and I'm willing to face any risk or fallout from that act.*
>
> No, don't '"bring it on'".
> The project needs to grow gradually so the software can be strengthened along the way.
> I make this appeal to WikiLeaks not to try to use Bitcoin. Bitcoin is a small beta community in its infancy. You would not stand to get more than pocket change, and the heat you would bring would likely destroy us at this stage.

I strongly agree with this sentiment, bitcoin is young, we should behave accordingly.

We should be working hard to grow and mature bitcoin with a goal that one day it will be of the standard and quality that it can be a great help for Wikileaks.

I suggest that when the bitcoin community commits to the sustained donation of more than 50% of Wikileaks operating expenses at a time to consider a formal relationship.

Until then, let's not use Wikileaks as a cheap publicity stunt, but rather a goal that we use to spur us on to develop this 'bitcoin' tool that will help support the important work that Wikileaks is doing on freedoms behalf.

Ps. In another post I voiced a differing view, which Wikileaks and bitcoin could enter into a mutually beneficial relationship for development; however I have since revised this view, under the account of the different maturities between Wikileaks and bitcoin. Wikileaks is mature, while bitcoin is (relatively) immature. It would not be a balanced relationship.

Bruce Wagner Writes:
Re: Wikileaks contact info?
December 05, 2010, 02:39:23 PM

Wow. Satoshi gets more than '"just one vote"'.... in my opinion.
Let's not let this issue divide us.

RHorning, and others, your commitment to what's right, freedom of the
press, etc is inspiring. It makes me want to support Wikileaks even more
than ever.

Also, our hero in Congress, Ron Paul, coming out supporting Wikileaks...
is very inspiring too. (Did you see this? Ron Paul - Julian Assange - Fed-
eral Reserve | http://goo.gl/NIBsk)

I propose a win-win solution to this.

Let's form a Wikileaks Support Task Force.... where we brainstorm about
ways to help Wikileaks ..... without bringing Bitcoin into the battle (or '"war"').

We are a large group of very smart guys here. We can probably put our
heads together and think of 1000 ways we could help Wikileaks .... even
in areas around funding, donations, etc.... WITHOUT recommending
Bitcoin. There are so many ideas we could give them. Even, in the area
of money, recommending other alternative payment systems ... Liberty
Reserve? I don't know... But you guys do! Recommend something far
better than PayPal --- super easy for Americans to use --- (but something
other than Bitcoin). MoneyBookers.com? GoldMoney.com? Paxum.
com? You tell me...

We can put our collective heads together and come up with 1000 great
ways we can HELP Wikileaks achieve what they need.... without mention-
ing Bitcoin.

Honestly, I'm not sure that Bitcoin really is their best alternate funding
channel anyway. It's NOT EASY for Americans to buy and send Bitcoin yet!

Just after my radio interview the other day, I spent 40 minutes on the
phone with a lady who called me from Toledo. She wanted me to talk

her through the entire process of creating an account at MyBitcoin, then buying Bitcoin on BitcoinGateway... It was excruciating. This woman had never used copy & paste before. Apparently, she never did fully complete the purchase verification process.

My point is:

1. Bitcoin is not ready for typical American users yet --- especially not millions of people all trying to use BitcoinGateway at once! Poor chaord will be pulling his hair out..... and probably taking his site & service off-line completely, as a result.

2. If we try not to be biased, we will see many other funding mechanisms that are probably much better suited for immediately use by average Americans, in USD, at this moment in time. Paxum?

3. Let's promote, propel, and support Wikileaks in every way.

4. Let's promote, propel, protect, and support our greatest tool of the 21st Century, Bitcoin, in every way. ....at the same time.

Bitcoin is our secret weapon in this '"war'". It is our Big Gun. Let's save it, preserve it, protect it, strengthen it...... and when the time is right, it will change everything.

If it gets too much bad publicity now... it's mainstream adoption could be delayed by DECADES. Seriously.

While they're busy trying to figure out how to block Wikileaks DNS ...

We are quietly getting all the world's most respected charities to accept Bitcoin. Then, all local barter clubs and local alternate currency groups, we convince to '"join the Bitcoin electronic network'"... Then, we get the world's most respected businesses to accept Bitcoin. All the while .... Bitcoin is getting stronger and stronger.... going more and more main-stream .... with every passing day. Under their radar. Right under their noses. In a way, it's using Bitcoin's obscurity to its best advantage... for this stage of its growth.

The giant Redwood ..... with a tiny ax... and a chop chop chop chop every day... it comes falling down.

Kiba Writes:
Re: Wikileaks contact info?
December 05, 2010, 05:33:39 PM

You do realize that time spent on helping wikileaks is not time spent on growing bitcoin. I think growing and making bitcoin better is a good use of our time. There are already ton of people helping wikileaks. When the time is right, bitcoin is going to change the world, just as it had changed how bitcoiners view money....forever.

Genjix
Re: Wikileaks contact info?
December 05, 2010, 07:29:10 PM

Having read this thread, I've done a U-turn on my earlier view and agree. Lets protect and care for bitcoin until she leaves her nursery onto the economic killing fields.

Kiba Writes:
Re: Wikileaks contact info?
December 05, 2010, 07:32:10 PM

bober182: If you're still here and you're still doing '"that thing"'. I suggest you turn around and tell the wikileaks folk that wikileak should stop considering bitcoin for the moment in case that Satoshi's message didn't get through.

Farmer_boy writes:
Re: Wikileaks contact info?
December 05, 2010, 09:16:16 PM

I am not the most stubborn person in the world. I side with Satoshi's wishes too.

ShadowOfHarbringer writes:
Re: Wikileaks contact info?
December 05, 2010, 09:19:43 PM

This is exactly what I was saying all along, but nobody is listening. Bitcoin is not yet strong enough to take on such a huge task.

Pushing bitcoin on wikileaks will kill bitcoin, and still it won't help wikileaks a lot.

RHorning writes:
Re: Wikileaks contact info?
December 05, 2010, 09:20:52 PM

I've stated my position on this, but I'm willing to listen to group consensus on this issue to so far as this isn't an irreversible decision. The fact is that there are dozens or hundreds of other opportunities that Bitcoins could help with, and I'm willing to work on most of those other ideas well before even trying to lobby Wikileaks for accepting Bitcoins as a donation.

If Wikileaks finds out and advertises the link for Bitcoin donations, we should be prepared for what might be a huge influx of new people into the forum and as users. I think the common consensus is that would happen even without a push by the Bitcoin community at all. Bitcoins are also low key enough that I personally don't think the popular news media (and with that the government regulators) would pay attention to Bitcoins on any sort of scale, but I can understand why people on this forum don't want to expose the Bitcoin community on this issue.

I still think it is the right thing to do and I personally am not afraid of what would happen to Bitcoin as software or to the network if such an influx of users and attention happened. Something is eventually going to break like that, so the Bitcoin community ought to be prepared for that eventuality. If this concrete example is somehow going to spur on developments that make Bitcoin more secure, it sounds like even the possibility that Wikileaks may use Bitcoins is a good thing. If not this particular incident, there will be something else like this which may eventually push Bitcoins into the limelight.

If Satoshi wants to stay timid, that is his opinion. We can't stay like this forever though.

ShadowofHarbringer writes:
Re: Wikileaks contact info?
December 05, 2010, 09:25:08 PM

Also, while bitcoin is in its infancy, we (bitcoiners) should avoid engaging in drugs, prostitution, weapon selling & other semi-legal activities.

If bitcoin will lift off using money from fraudesters & gangsters, how in world will we ever manage to promote it to be a serious currency ?

FatherMcGruder
Re: Wikileaks contact info?
December 07, 2010, 04:09:30 AM
It really isn't up to us whether or not Wikileaks adopts bitcoin. Isn't that part of the currency's beauty?

Bruce Wagner writes:
Re: Wikileaks contact info?
December 11, 2010, 06:56:03 PM
I guess it's too late now.
The words bitcoin and wikileaks have appeared in the same headline in pc world just now.
Search google news for the word bitcoin now.
[https://www.pcworld.com/article/213230/could_wikileaks_scandal_lead_to_new_virtual_currency.html]

The *PC World* article on Bitcoin.

Outlines the possibility of Wikileaks using Bitcoin following the Paypal financial blockade of the organisation.

S3052 writes:
Re: Wikileaks contact info?
December 11, 2010, 07:28:31 PM
Agree Bruce.

This is not necessarily negative, as long as we use the momentum behind that in a positive way - and stay out of the direct connection to wikileaks.

We should stay distinctively distant from wikileaks, meaning: We should not push wikileaks to accept donations in bitcoins at all.

ShadowOfHarbringer
Re: Wikileaks contact info?
December 12, 2010, 01:30:34 AM
As you said, the genie is out of the bottle. It is all up to wikileaks now.
I sincerelysincererly hope they don't adopt bitcoin.

CW writes:
Re: Wikileaks contact info?
April 02, 2011, 07:00:09 PM
Quote from: S3052 on December 11, 2010, 07:43:38 AM
The art has begun now to turn this new bitcoin publicity into positive and legitimate activities, ideally even stating that bitcoin does not want to be assiciated with wikileaks af this stage (like satoshi has expressed clearly) we could add a statement on the homepage of bitcoin.org.
Please, don't do that. Wikileaks might have some bad press in USA, but they're seen as heroes in the rest of the world (for example, a recent survey shown recently that Julian Assange is the third most reputated '"international leader'" for Spanish people). Just be neutral, it's not bitcoin's role to endorse or condemn Wikileaks. If bitcoin.org condemns Wikieaks, it'll be EPIC FAIL for bitcoin reputation in many places all over the world.

Shadow of Harbringer writes:
Re: Wikileaks contact info?
April 02, 2011, 09:24:16 PM
+1

Currency should stay a currency. When we involve in political stuff, we risk being associated with one of the sides.
Currencies should be neutral by design.

The last public post made by Satoshi Nakamoto.

Source: Bitcointalk

Satoshi writes:

Added some DoS limits, removed safe mode (0.3.19)

December 12, 2010, 06:22:33 PM

There's more work to do on DoS, but I'm doing a quick build of what I have so far in case it's needed, before venturing into more complex ideas. The build for this is version 0.3.19.

- Added some DoS controls

As Gavin and I have said clearly before, the software is not at all resistant to DoS attack. This is one improvement, but there are still more ways to attack than I can count.

I'm leaving the -limitfreerelay part as a switch for now and it's there if you need it.

- Removed '"safe mode"' alerts

'"safe mode"' alerts was a temporary measure after the 0.3.9 overflow bug. We can say all we want that users can just run with '"-disablesafemode"', but it's better just not to have it for the sake of appearances. It was never intended as a long term feature. Safe mode can still be triggered by seeing a longer (greater total PoW) invalid block chain.

Builds:

http://sourceforge.net/projects/bitcoin/files/Bitcoin/bitcoin-0.3.19/

The infamous 'I've moved on to other things' email.

In an email exchange with core Bitcoin developer Mike Hearn, Satoshi casually mentions his retreat.

Source: https://pastebin.com/syrmi3ET

From: Satoshi Nakamoto <satoshin@gmx.com>
Date: Sat, Apr 23, 2011 at 3:40 PM
To: Mike Hearn <mike@plan99.net>

I had a few other things on my mind (as always). One is, are you planning on rejoining the community at some point (eg for code reviews), or is your plan to permanently step back from the limelight?

I've moved on to other things. It's in good hands with Gavin and everyone.

I do hope your BitcoinJ continues to be developed into an alternative client. It gives Java devs something to work on, and it's easier with a simpler foundation that doesn't have to do everything. It'll get critical mass when impatient new users can get started using it while the other one is still downloading the blockchain.

Newsweek Dorian Satoshi Nakamoto Claim.

After *Newsweek* publish an article claiming that Dorian Satoshi Nakamoto, a middle-aged Japanese American living in California was the 'real' Satoshi, Satoshi breaks a three-year silence to make a public post on the P2P Foundation's website.

P2P Foundation
2014-03-06 00:00:00 UTC
  I am not Dorian Nakamoto.

March 7th, 2014 Newsweek article:
http://www.newsweek.com/2014/03/14/face-behind-bitcoin-247957.html

On PGP and book bans in 1994.

This email exchange constitutes a short excerpt from the *Cypherpunks Mailing list* in 1994. It is included here due to its historical relevance to the early Crypto Wars discussed in James Bridle's introduction and because Hal Finney was notably the first person to ever receive a Bitcoin transaction from Satoshi Nakamoto.

1994-01-02 - POLI: Politics vs Technology

Header Data
From: Hal <hfinney@shell.portal.com>
To: cypherpunks@toad.com
Message Hash: adf1a77235794be59d5f3f3c4f983f0f0a6cc6bee27491b-d8e645bb1755220c7
Message ID: <199401021857.KAA16654@jobe.shell.portal.com>
Reply To: N/A
UTC Datetime: 1994-01-02 18:58:40 UTC
Raw Date: Sun, 2 Jan 94 10:58:40 PST

Raw message
From: Hal <hfinney@shell.portal.com>
Date: Sun, 2 Jan 94 10:58:40 PST
To: cypherpunks@toad.com
Subject: POLI: Politics vs Technology
Message-ID: <199401021857.KAA16654@jobe.shell.portal.com>
MIME-Version: 1.0
Content-Type: text/plain

From: Mike Ingle <MIKEINGLE@delphi.com>
> But could the government ban a book today? Of course
> not, at least not after one person typed it or scanned
> it into a computer. Technological gains are permanent.
> The political approach is only useful as a tactical
> weapon, to hold them off until technological solutions

>are in place. If you want to change the world, don't
>protest. Write code!

This position seems to be fast becoming Cypherpunks
dogma, but I don't agree. The notion that we can
just fade into cypherspace and ignore the unpleasant
political realities is unrealistic, in my view.

Have people forgotten the Clipper proposal, with the
possible follow-on to make non-Clipper encryption
illegal? To the extent this proposal has been or
will be defeated, it will happen through political
maneuvering, not technology.

Have people forgotten the PGP export investigation?
Phil Zimmermann hasn't. He and others may be facing
the prospect of ten years in prison if they were found
guilty of illegal export. If anyone has any suggestions
for how to escape from jail into cyberspace I'd like to
hear about them.

Mike's SecureDrive is a terrific program for protecting
privacy. But if we want to keep keys secret from
politically-motivated investigations, we have to rely
on the very political and non-technological Fifth
Amendment (an amendment which Mike Godwin of EFF and
others contend does not actually protect disclosure of
cryptographic keys). Again, we need to win political,
not technological, victories in order to protect our
privacy.

I even question Mike's point about the government's
inability to ban books. Look at the difficulty in keeping
PGP available in this country even though it is legal.
Not only have FTP sites been steadily closed down,
even the key servers have as well. And this is legal
software.

Sure, this software is currently available overseas, but that is because PGP's only legal limitations are the U.S. patent issues. Imagine how much worse it would be if non-escrowed encryption were made illegal in a broad range of countries, with stringent limits on net access to countries which promote illegal software? Here again, these kinds of decisions will be made in the political realm.

Fundamentally, I believe we will have the kind of society that most people want. If we want freedom and privacy, we must persuade others that these are worth having. There are no shortcuts. Withdrawing into technology is like pulling the blankets over your head. It feels good for a while, until reality catches up. The next Clipper or Digital Telephony proposal will provide a rude awakening.

Hal Finney
hfinney@shell.portal.com

http://mailing-list-archive.cryptoanarchy.wiki/archive/1994/01/ad-f1a77235794be59d5f3f3c4f983f0f0a6cc6bee27491bd8e645bb1755

Acknowledgements

Many thanks to the following individuals, without whose contributions this book would not have been possible. Eva Jäger for her in-depth research and early conversations on the development of the book. Cecilia Serafini, our designer and saviour. Heather Parry for her patient and meticulous copyediting of the text. Jay Springett for his ongoing guidance, collaboration and taking the time to feedback on the early stages of this book. Sarah Shin for her grace, patience and for applying the appropriate level of pressure to ensure this book actually saw the light of day.

Special thanks to all those that have contributed either through friendship, collaboration or in their insight into the developments of blockchain technologies, including (but likely not limited too) Ruth Catlow, Takeshi Shiomitsu, Elias Haase, Sam Hart, Rob Myers, The Wine and Cheese Appreciation Society of Greater London, Nick Szabo, Amir Taaki, those present in the Crypto Circle, the Satoshi Nakamoto Institute, Guild, the cryptorave scene and all those who, for good reason, shall remain forever nameless.

And an eternal thanks to Jake Vickers for his patience and dedication to sharing the complicated task of constructing and maintaining cryptocurrency mining rigs, which in more ways than one assisted in the development of this book.

Biographies

Jaya Klara Brekke writes, does research and speaks on the political economy of blockchain and consensus protocols, focusing on questions of politics, redistribution and power in distributed systems. She is the author of the B9Lab ethical training module for blockchain developers, and the Satoshi Oath, a hippocratic oath for blockchain development. She is writing a PhD with the preliminary title Distributing Chains, three strategies for thinking blockchain politically (distributingchains.info).

James Bridle is an artist and writer working across technologies and disciplines. His artworks have been commissioned by galleries and institutions and exhibited worldwide and on the internet. His writing on literature, culture and networks has appeared in publications including *Wired*, *Domus*, *Cabinet*, the *Atlantic*, the *New Statesman*, the *Guardian*, the *Observer* and many others. *New Dark Age*, his book about technology, knowledge, and the end of the future, was published by Verso in 2018. His work can be found at http://jamesbridle.com.

Satoshi Nakamoto is the identity used by the unknown person or people who developed Bitcoin, authored the Bitcoin white paper, and created and deployed the first Bitcoin implementation. As part of the implementation, they also devised the first blockchain database. Satoshi Nakamoto ceased public involvement with Bitcoin at the end of 2010; their last public post was made in 2014 as a rebuttal to claims on the 'true' nature of Satoshi's identity. During Bitcoin's peak in December 2017, the Satoshi Nakamoto identity could lay claim to a fortune worth over $19 billion, making Nakamoto possibly the forty-fourth-richest person in the world at that time. To this day both the public identity and Bitcoin wallet attached to the identity of Satoshi Nakamoto remain inactive.

Ben Vickers is a curator, writer, explorer, publisher, technologist and luddite. He is CTO at the Serpentine Galleries in London, co-founder of Ignota Books and an initiator of the open-source monastic order unMonastery.